12-22-76

35: International Broadcasting: A New Dimension of Western Diplomacy

THE WASHINGTON PAPERS
Volume IV

35: International Broadcasting:
A New Dimension
of Western Diplomacy

David M. Abshire

THE CENTER FOR STRATEGIC AND INTERNATIONAL STUDIES
Georgetown University, Washington, D.C.

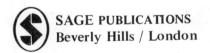

SAGE PUBLICATIONS
Beverly Hills / London

For information address:

SAGE PUBLICATIONS, INC.
275 South Beverly Drive
Beverly Hills, California 902 12

SAGE PUBLICATIONS LTD
St George's House / 44 Hatton Garden
London EC1N 8ER

International Standard Book Number 0-8039-0657-9

Library of Congress Catalog Card No. 76-21919

SECOND PRINTING

*When citing a Washington Paper, please use the proper form. Remember to cite
the series title and include the paper number. One of the two following formats
can be adapted (depending on the style manual used):*

(1) HASSNER, P. (1973) "Europe in the Age of Negotiation." The Washington
Papers, I, 8. Beverly Hills and London: Sage Pubns.

OR

(2) Hassner, Pierre. 1973. *Europe in the Age of Negotiation.* The Washington
Papers, vol. 1, no. 8. Beverly Hills and London: Sage Publications.

1953297

CONTENTS

Foreword by Foy D. Kohler   7

Author's Preface   8

I. Helsinki and Beyond   9

II. The Political Roles of International Broadcasting   17

     The National Services   17

     The Surrogate Home Services   28

III. Soviet Control of Communications   40

IV. The Soviet Union: Ferment, Repression, and Change   53

V. International Broadcasting to Eastern Europe   65

VI. Toward a New Communications Diplomacy   76

     References   88

FOREWORD

Americans and other Westerners can hardly conceive of life in a society where all information media are a monopoly of the ruling regime, as in the Soviet Union and the countries of Eastern Europe. One has to live there, as I did for more than seven of the past 40 years, to realize with what intense eagerness one turns constantly to radio broadcasts from abroad to learn what the rest of the world is saying and thinking, even what is going on in the country itself. The Russians, the Ukrainians, the Poles and the other people of the so-called "Socialist Commonwealth" share this feeling and fortunately are able to do the same, for foreign radio transmission is the principal medium which escapes the total monopoly. With vast distances to cover, these regimes have long relied on short-wave broadcasting for their own indoctrination purposes and have made short-wave receivers one of the few consumer items relatively easy to obtain. They have thus provided a channel for communication between the peoples of East and West, which the Soviets would like to block out, but are unable to do effectively.

The importance of maintaining this free radio channel of communication and of further developing this and other means of reaching Soviet and Eastern European peoples can hardly be exaggerated. Western efforts in this direction have visibly contributed to the development of an element of public opinion in these societies that did not exist in Stalin's day, and this can help to promote an evolutionary process that will eventually make it easier for the West to live with them in peace.

Since in our country an informed public opinion is the indispensable basis for all policy and action, I should like to see every American become familiar with Dr. Abshire's study of international broadcasting as an instrument of Western diplomacy and his persuasive analysis of the problems and prospects which face us in this field today.

Foy D. Kohler
U.S. Ambassador to the Soviet Union (1962-1966)
Former Director, Voice of America

Author's Preface

Appearances before Congress to testify on international broadcasting, and discussions with Western diplomats and broadcasters, convinced me of the pressing need for an overall analysis of its emerging significance and role in Western diplomacy. Finally I myself undertook such a study. After my first draft, many experts contributed in various ways to the final production. A special debt of gratitude goes to four friends who worked with me throughout on the improvement of the manuscript: Gayle Durham Hannah, Dimitri Simes, I. R. Wechsler, and Dean Guntner White. In addition, I was helped by many knowledgeable persons, including Harold Anderson, William Griffith, Jan Karski, Stanley Leinwoll, Alexander Lieven, Francis Murray, Francis Ronalds, Anatole Shub, and Ralph Walter. This Washington Paper is part of the CSIS program on International Communications and Cultural Relations which developed while Walter Roberts was at CSIS, and to him I am indebted for additional inspiration in the larger field of information and cultural affairs.

To Walter Laqueur, Editor, and Ethel Eanet, Managing Editor of the Washington Papers, my thanks for their particular attention in editing my manuscript and seeing it through its publication. To Elise Thoren, my long-time friend, who patiently typed many versions of the draft manuscript, my sincere gratitude.

Finally, it goes without saying that the views herein are my own, and do not necessarily reflect those of the institutions and boards with which I have been associated, or the experts who have aided me in this study.

—David M. Abshire

I. HELSINKI AND BEYOND

For three decades the prospects for a more peaceful and stable world have been dominated by the relationship between the United States and the Soviet Union. Moving from the grim hostility of Stalin's day, both nations in the present era declare it their policy to achieve a more constructive and cooperative relationship. However, leaving aside other currently-debated aspects and issues of détente, it is clear that the proclaimed relaxation of tension in the Soviet Union's foreign relationships has brought no relaxation of its internal controls. With each assertion of dedication to peaceful coexistence has gone a reassertion of the need for intensified ideological struggle and control of the ideas and information of the Soviet peoples.

Thus there is an uneasy dichotomy in Soviet behavior, which imposes a dangerous duality in perceptions of the world. Without a common body of information there can be no common understanding, no shared appreciation of actuality, and no firm base for a more stable relationship. From this perspective, broadcasting to communist societies takes on a new role and a new responsibility. Soviet relaxation of stringent control and censorship does not seem a present prospect—either as a consequence of détente or as a support for it. But international broadcasting, unlike all the other media of communication—books, works of art, newspapers and magazines, and even word-of-mouth—cannot be stopped at frontiers. It comes into a listener's home always as an invited guest, admitted or rejected by the turn of a dial; and it cannot be selectively edited or censored by an intermediary. It alone can sustain the dialogue required for progress toward a broadened and more constructive level of coexistence.

The significance of international broadcasting is not that it can say things over greater distances, or say them to more people, or say them more loudly. Each new medium of communication has brought new possibilities for the creation of human community, expanding the common store of human knowledge, and of peoples' knowledge of each other and of themselves. The importance of international broadcasting is that it offers a major channel for establishing that communication between nations and peoples necessary to a reliable structure of peace. As Wolfgang Leonhard (1973: 74) has noted:

> The normalization of relations between East and West cannot be limited to treaties between governments; to be real and lasting, it should include relations between peoples as well. Therein lies the true test: East-West rapprochement should not be measured by the number of summit conferences and the extent of trade agreements, but above all by the degree to which there is a free exchange of ideas and culture between the peoples of the two areas.

The Pledges of Helsinki

In the summer of 1975, after two years of negotiation, there was a major summit conference—the Conference on Security and Cooperation in Europe (CSCE)—which was highly significant for the future of international broadcasting to Eastern Europe and the Soviet Union. The final Helsinki summit of 35 heads of state produced a 120-page document; its major concerns were security in Europe and in the Mediterranean; economic, scientific, technical and environmental cooperation; and cooperation in humanitarian and other fields including information (a group of issues that became known as "Basket III"). The section on human contacts and information declared that the signatories were conscious of the need for greater mutual knowledge and understanding of life in its various aspects among their several countries.

The signatories acknowledged the contribution of this process "to the growth of confidence among peoples" and said they desired continuing efforts towards developing mutual understanding and improvement of relations among themselves. They recognized the importance of mutual dissemination of information and emphasized "the essential and influential role of the press, radio, television, cinema and news agencies and of the journalists working in these fields." The aim of all these affirmations was to further the free availability of information of all kinds. Regarding broadcast information, the document was specific:

> The participating states note the expansion in the dissemination of information broadcast by radio, and express the hope for the continuation of this process, so as to meet the interest of mutual understanding among peoples and the aims set forth by this Conference [U.S. Dept. of State, 1975: 117-118].

To the Western participants, this statement was clearly an approving reference to the decrease in communist jamming since 1973, recognizing it as mutually beneficial that the communist countries no longer jam the British Broadcasting Corporation (BBC), the Voice of America (VOA), or Deutsche Welle (DW). United States negotiators had attempted to insert explicit references to the cessation of jamming. Predictably, the effort failed. The Soviet representatives refused to accept precise language and insisted on ambiguous formulas. While declaring its commitment to freedom of information, the Soviet government eluded firm adherence to specific policies. However, in expressing hope for the continued "expansion in the dissemination of information by radio," the final act implies that jamming of other broadcasts (Radio Liberty, Radio Free Europe, Kol Israel, Radio Peking) should cease (Board for International Broadcasting [BIB], 1975: 1).

Soviet Motives at Helsinki

From the early 1950s onward, the proposed conference on European Security and Cooperation had assumed great importance to the Kremlin. The Soviet Union and its Warsaw Pact allies had steadily pressed for scuh a meeting, eventually showing a willingness to make some concessions to allow it to take place. Originally, the Soviet design was to involve only European states, but West European countries insisted on participation by the

United States and Canada. Consequently, in 1970 the Kremlin modified its proposal to include all North Atlantic Treaty Organization (NATO) members, and accepted NATO preconditions that the Berlin issue be resolved (as it was in 1972) and that preliminary talks begin on mutual balanced force reductions (MBFR) in Europe. Initial stages of the conference were held in Geneva in 1973.

Leonard Brezhnev had his own special motivation for this conference. At the 24th Party Congress in 1971 he committed himself to it as part of the party's "peace program." Among the tasks ahead, he said, was "to proceed from the final recognition of the territorial changes that took place in Europe as a result of the Second World War." By 1974 he was calling for the conference to terminate with a summit meeting of chiefs of state, so dramatizing his achievement prior to the 25th Communist Party Congress planned for early 1976.

In the West, much of the original justification for agreeing to a conference, which the Kremlin saw as ratifying the status quo, was an expected advance towards freer movement of people, ideas, and information. NATO had shown that it was not seeking to impose a change of the status quo, and this, it was hoped, would elicit an improvement in Soviet treatment on human rights. If present borders were stabilized, those borders should be opened to an increasing flow of communication and human exchanges. At the emphatic insistence of Western European nations and Yugoslavia and Finland, the section on humanitarian cooperation—"Basket III"—was added.

To Brezhnev and his Politburo colleagues, the conference declaration on the "inviolability of frontiers" and "nonintervention in internal affairs of other nations" completed a historical process begun at Yalta. The long-sought conference, in the Soviet view, had legitimized Soviet hegemony over Eastern Europe. The leadership also sought another advantage from the conference: an improvement, at low cost, of the Soviet image in the West by presenting Soviet foreign policy as moderate and cooperative. They felt this gain could be achieved without substantial concessions in opening the communist nations to "subversive" information from the West.

In the compromises of the negotiations leading to the Final Act, the Soviet Union, eager to obtain its objectives, did make some concessions. In "Basket Three" there were provisions on family reunification and visits, binational marriages, improved working conditions for journalists, and increased cultural and educational exchanges.

The Western leaders, especially Ford and Kissinger, have emphasized that the declaration was only that—neither a treaty nor legally binding; that we did not consider it an additional ratification of any existing agreement; that it did not put a seal of approval on the existing divisions of Europe; and that the doctrine of inviolability of frontiers and non-interference in internal affairs cut both ways, and could—by Western interpretations—apply to potential Soviet actions with regard to Romania, Yugoslavia, and Portugal.

Although the signatories pledged themselves to make information more freely available, specific stipulations were often hedged. Dissemination of "oral information" was to be through lectures and lecture tours, exchange of opinions at round-table meetings, summer schools, congresses, and other bilateral and multilateral meetings—all very carefully controlled events. They pledged improved dissemination of newspapers and other publications, but with the qualification that for this purpose "competent firms and organizations" would be encouraged to conclude agreements designed to "gradually" increase the number and titles imported.

Little was achieved to ensure free access to this information by the peoples involved, and this, of course, goes to the heart of the censorship problem. The Soviet Union, for example, rejected a Western proposal for the right to set up reading rooms. But by hard bargaining and unanimity Western diplomats forced the inclusion of the human rights section. Its provisions—which the Soviet Union had never wanted in the agreement—touched a vital nerve in the Soviet system of information control.

Preserving the Barriers

Both before and after the conference, Soviet leaders repeatedly demonstrated their sensitivity to the freer flow of information. In speaking to the Soviet Trade Union Congress in 1972, Brezhnev had called for "implacability in the ideological struggle combined with readiness to develop advantageous relations with states of the opposite social system." The leading ideologist of the Politburo, Mikhail Suslov, looking ahead to the conference, emphasized that

> In all our ideological work no relaxation of the struggle against reactionary bourgeois ideology is admissible. . . . Precisely in the field of ideology, the field where there is not and cannot be any peaceful coexistence, the struggle has been sharpened considerably [Report, 1973: 8-9].

In June of 1975, while the Soviet Union was negotiating on "Basket III," KGB Chief Yuri Andropov in a rare public statement accused "socialism's adversaries," in the guise of "democratizing socialism" and defending civil rights in the Soviet Union, of seeking to undermine Soviet power from within.

Nor surprisingly, the Helsinki declaration was followed by an increase of such Soviet statements and efforts to impose limits and qualifications on it. There were pressures on the Soviet Union. The first of these was that leaders of the democratic West, notably Harold Wilson and Valery Giscard d'Estaing, warned Moscow that its commitment to détente would largely be judged by its response to the section on human rights. Giscard even called upon the Soviet Union to participate in an ideological détente. Gerald Ford stressed at Helsinki that history would judge the conference "not by the promises we make, but by the promises we keep."

A second important pressure on the Soviet Union was that many Soviet dissidents began to cite the language of the human rights section in their struggle for greater justice at home. This pertinence of "Basket III" to the cause of the Soviet dissidents, and its use in their own protest movement, was apparently never fully anticipated by the Soviet negotiators (Larrabee, 1976).

The ink was hardly dry on the Helsinki document when the Soviet Union began a concerted campaign to impose its own interpretations. A leading spokesman in this effort was Georgiy Arbatov, head of the Institute for the Study of the United States of America and Canada. According to Arbatov, if "some people regard" provisions of the Helsinki final document as "an invitation to fling open the door to subversive anti-Soviet pro-violence propaganda, or to fan national, and racial strife, then they are laboring in vain. Neither the document signed in Helsinki nor détente will permit such occurrences." He asserted that in establishing diplomatic relations in 1933 the United States and the Soviet Union agreed not to permit interference in each other's internal affairs and that this agreement could not be squared "with the subversive activities conducted by the radio stations Radio Liberty and Radio Free Europe" (*New York Times,* Oct. 8, 1975). In other words the Soviet Union, adhering to Leninist communications doctrine, intends to maintain barriers against any information and ideas contrary to the official line.

As part of this campaign the Soviet Union and some of its East European clients have begun new propaganda attacks against Western broadcasts, especially Radio Liberty and Radio Free Europe. *Pravda,* in a major article in January 1976, charged Radio Liberty and Radio Free Europe with trying to change the communist system in the Soviet Union and Eastern Europe. The article claimed that the "international public" was indignant because "official U.S. institutions direct and finance this subversive activity." The activity of the radio stations was said to be "incompatible with the final act of the pan-European conference [Helsinki], with the elementary norms of international law, with the process of lessening international tensions." Other articles accused the Voice of America of not reflecting the détente policy of the administration (Pond, 1976).

In further contrast to the Helsinki agreement, communist representatives pressed the International Olympic Committee into revoking Radio Free Europe's accreditation at the 1976 Innsbruck Olympic Games, and prevented the West German radios Deutsche Welle and Deutschlandfunk reporting from the March 1976 Leipzig Fair. The BBC interview with Solzhenitsyn, televised that same month, resulted in Soviet denial of a visa to Sir Charles Curran, director-general of the BBC.

Plainly, peaceful coexistence is not to mean peaceful ideological détente, as Giscard d'Estaing had urged. Instead, for the Soviet Union it means "strengthening the struggle against bourgeoisie ideology" and greater "intolerance" towards Western ideas, in the words of Politburo member Vitaliy Grishin to the 25th Party Congress (*Pravda,* Feb. 26, 1976). At that Party Congress, chief ideologist Mikhail Suslov received special prominence and several officials responsible for Soviet propaganda were promoted to full central committee membership, a clear indication of growing Soviet attention to ideological conflict with the West.

Only in the area of international broadcasting is the Soviet Union unable to maintain information control. Thus the dichotomy in Soviet détente policy gives international broadcasting a new significance and a new task, if the Helsinki commitment to freer movement of information and ideas is to be meaningful. And a more coherent and vigorous international broadcasting diplomacy on the part of the Western democracies is needed. To judge the performance of international broadcasting, and its potentialities for assuring that peaceful relationships rest on some basis of freely shared facts and ideas, requires that we examine the history and present patterns of international broadcasting, the Soviet perspective on communication, the situation and responses of listeners, and the problems and possibilities that lie ahead. The focus in this work will be on the Soviet Union and its Eastern European allies; the Soviet Union presents the pivotal and crucial case, and it is in these areas that international broadcasting has its widest experience and access.

II. THE POLITICAL ROLES OF INTERNATIONAL BROADCASTING

THE NATIONAL SERVICES

It is not easy for Americans to appreciate the dimensions or impact of international broadcasting. We live in a dense network of communication facilities of every kind. Geographic mobility is one of the most salient features of our society, and we have easy and prompt access to news, views, and information of all kinds—indeed, we sometimes feel that we suffer from a surfeit. Only two countries touch our borders. We rely primarily on our own media to bring us accounts of developments abroad and we are not in the habit of listening to foreign radio. We do not need and have not developed an extensive short-wave radio network to cover vast distances.

However, "radio remains the main mass medium of communication in two-thirds of the world" (BBC, 1974: 267). The transistor revolution made radio a major medium almost everywhere. For example, in Latin America the number of sets has risen from 12 million in 1955 to over 60 million today; in India, from a million to over 20 million; in the countries of black Africa from 360,000 to over 20 million (BBC Handbook, 1976: 54). In the Soviet Union and the communist countries of Eastern Europe, ownership of radio receivers increased more than four-fold from 1955 to 1974 when sets numbered approximately 90 million (BBC Handbook, 1976: 54; World, 1976; 466-468). The proportion able to receive short-wave is believed to be about two-thirds.

The communications explosion—with its ability to transport ideas and information to millions of people simultaneously and instantaneously—is bringing about a many-sided revolution in much of the world, ending rural isolation, and changing traditional attitudes. And radio is serving worldwide as a vehicle for entertainment and for the dissemination of serious and popular cultural productions at all levels.

Pre-World War I radio amateurs pioneered in the field of long distance broadcasting, experimenting with progressively shorter wave lengths in their search for effective communication across national boundaries. But it was primarily political motivation that brought about the expansion of international broadcasting, which having accelerated explosively, is still continuing to grow.

Technology was still too primitive, and the number of receivers too few, for radio to play much of a role in World War I. Germany, its international cables cut, initiated a news service to neutrals in 1915, which may have been the first international use of radio for political purposes. In Russia, the revolutionary leadership was quick to see possibilities for reaching both domestic and foreign audiences. The cruiser *Aurora,* in the harbor of what was then Petrograd, broadcast messages from Lenin to "the citizens of Russia" on the morning of the Bolshevik seizure of power on November 7, 1917 (Paulu, 1974: 33). They also used radio to send their ideological message abroad, to justify their separate peace with Germany, and to support Trotsky's maneuvers in negotiating the Treaty of Brest-Litovsk. In 1922 Lenin, who had described radio as "a newspaper without paper . . . and without boundaries," established in Moscow what was then the most powerful broadcasting station in the world.

Between the two world wars, the political use of radio expanded. France and Germany engaged in a battle of words over the airwaves when the Ruhr was occupied in 1923, and the Soviet Union's revolutionary propaganda directed against the Romanian province of Bessarabia (now Soviet Moldavia) led to a jamming effort—apparently the first—by Bucharest. Although the Italians used short-wave broadcasts in the Ethiopian War, the Spanish Civil War brought its first extensive use by both sides in wartime.

The Axis powers, like the Soviet Union, were vigorous in seizing and exploiting the opportunities radio offered. They used it for ruthless and insistent indoctrination at home, and for psychological warfare overseas, preparing for the aggressions of World War II. The democracies were slow to respond to the Axis powers' radio "war of nerves." Thus, "as late as 1937, while Moscow was broadcasting in seven languages, Rome in sixteen, and Berlin in six, the British were still content to use English only" (Whitton and Larson, 1964: 37).

The first interest in short-wave broadcasts of the European democracies arose from their need to maintain and strengthen the links with their overseas possessions. Holland began a service to the Dutch East Indies in 1927, France began French broadcasts to her colonies in 1931, and the British initiated the BBC Empire Service in 1932. British broadcasts in foreign languages (Arabic, French, Italian, and German) were started only in 1938. In the United States, there had been some private commercial and religious short-wave broadcasting—mainly to Latin America—but not until February 1942 did the United States begin government overseas broadcasts.

International Broadcasting in World War II

The outbreak of war in 1939 gave new dimensions and impetus to international broadcasting, and a new importance to public opinion. In most previous wars, belligerents (and neutrals, too) could exercise censorship and control over information entering their own country. Messages had to be physically transported across borders, and the general public could learn about developments inconsistent with official versions only slowly and with difficulty, if at all. International radio ended a nation's ability to insulate its population. Popular views and opinions, morale and will, could now be direct targets on a grand scale; for offense and defense, the term "home front" took on new significance.

Under Hitler and Goebbels, radio became an unprecedented instrument of psychological warfare. Nazi propaganda from the first was carefully orchestrated and unscrupulous. It entered the war fully equipped and staffed. "German radio propaganda to foreign countries had been 'at war' ever since the Party came to power" (Fraser, 1957: 74).

German external broadcasts selectively slanted their content to audiences, and cynically mixed truth and falsehood. They concentrated on short-term gains rather than the long term. They expected a quick victory—and said so. The same techniques that had served to bring Hitler to power were counted on to succeed with foreign audiences; perhaps the salient weakness of German propaganda was its inability to achieve any direct and imaginative identification with foreign states of mind. It was skillful in techniques of hysteria, hate, and terror, and formidable in times of victory.

Some of the most effective days of Nazi propaganda broadcasting were at the heights of their successes, especially in the spring of 1940 with the successful invasion of France. The French, suspicious and disillusioned, listened avidly to German radio, its effect enhanced by the low credibility of the French press and radio among its own people. As the Nazis began to suffer reverses, however, the inconsistencies of their propaganda became apparent and it was no longer believed. Goebbels increaingly gave orders to slant the news, asserting, in the spring of 1942: "News is a weapon of war. Its purpose is to wage war and not to give out information" (Hale, 1975: 10).

The wartime radio operations of the Western allies pursued a different strategy, although the tactical use of broadcasts for psychological support of military operations was highly developed and successful. The BBC External Service capitalized on the Empire Service initiated in 1932, continuing and consolidating the principles established by Sir John Reith, the first head of the BBC. He had from the beginning insisted on accuracy, and had been concerned to maintain an organizational structure and climate independent of the government in power. With the outbreak of war, some questioned whether the BBC should continue to place reliability and accuracy at the head of its priorities. The original policy was sustained, however. The then-director of the German-language services, Sir Hugh Greene (1969: 21), outlined its operating principles:

> To tell the truth within the limits of the information at our disposal and to tell it consistently and frankly. This involved a determination never to play down a disaster. . . . Then our audience in Germany and in the German forces, having heard us talking frankly about our defeats, will believe us when we talk about our victories, and the will to resist in a hopeless situation could, one hoped, be effectively undermined.

The United States, on being forced into the war, took the same position. The first broadcast of the short-wave Voice of America (VOA) on February 24, 1942 began:

> Today America has been at war for seventy-nine days. Daily we shall speak to you at this time about America and the war. The news may be good or bad. We shall tell you the truth [Dizard, 1961: 33].

VOA, under the Office of War Information (OWI), in spite of its inexperience and inevitable clashes within the wartime bureaucracy about its role and emphases, rapidly expanded both its competence and facilities. It put strong emphasis on prompt news and its interpretation, assuming that most audiences in enemy-controlled areas lacked uncensored news and background, and enlisted many of its staff from journalism. It also had to recruit heavily from emigré skills; its leadership, however, insisted on VOA speaking from an American position rather than as a reflection of emigré factions. Its programming was more directly and elaborately supervised than the BBC's. By 1945, from the 13 available short-wave transmitters originally taken over, VOA had 36 transmitters in the continental United States, and 14 installed overseas, plus captured facilities in Europe and Asia (Thompson, 1948: 43-44).

International broadcasting had come of age and was to be a key factor in foreign relations in the future. Even before the war ended, belligerents and most neutrals had virtually covered the globe, broadcasting in over 40 languages from more than 340 transmitters in 55 countries (Whitton and Larson, 1964: 44).

Postwar Broadcasting: The Soviet Union

When hostilities ended, the Western democracies largely demobilized their wartime propaganda organizations, as they did their military forces. The Soviet Union, on the other hand, preserved and enlarged all its instruments of international leverage, both state and party, to launch an aggressive campaign of pressure and expansion. Radio Moscow had been highly active during the the war. Its basic themes were "denigrating the Axis and celebrating Russia's lone role in the War" (Thompson, 1948: 99). The

Soviets had made ideological emphasis on revolution and class struggle subordinate to basic patriotic motifs, but attempted little cooperation with Allied propaganda, which gave considerable prominence to Russian wartime achievements.

After the war the Soviets expanded their foreign broadcasting, with a four-fold increase from 1948 to 1972, half of it from 1959 to 1969 (Report, 1973: 14-20). In August 1964, Radio Moscow was joined by a new international station. Designated "Radio Peace and Progress," it was presented as an independent "voice of the people," supposedly divorced from government in the manner of the BBC. This was clearly a response to the effectiveness of foreign broadcasting from the West. The change in programming was marked by a far more aggressive propaganda content.

The Soviet Union and Eastern Europe have taken full advantage of the political freedom of North America and Western Europe, directing some 1,400 hours of broadcasting a week there, 164 hours of that total in English to North America (Report, 1973: 14). International broadcasts from Warsaw Pact countries have doubled in the last 20 years and are growing rapidly; these alone now total over 1,400 hours per week (BBC Handbook, 1976: 65).

In 1976 the Soviet Union's broadcasts to foreign audiences totalled nearly 2,000 hours a week in 84 languages (BBC Handbook, 1976: 65), almost equal to the combined hours of VOA, RL, and RFE. In addition, Radio Moscow home service now transmits 190 hours a week to Soviet citizens abroad. About 11 percent of Soviet external broadcasts, or about 221 hours, are directed to the Western hemisphere, including 109 hours to Latin America. Almost 10 percent is beamed to India—171 hours in 13 languages. Far East transmissions rose to 333 hours, and the Near East received 186 hours. Broadcasts to black Africa total 168 hours.

West European Broadcasting: The BBC, Deutsche Welle

The development of the major broadcasters to the Soviet Union and Eastern Europe has marked in no small degree the state of the East-West relationship. In the diverse patterns of their organizational structure, concepts of mission, emphases and practices, these international broadcasters have also reflected their countries' varied international roles.

After the war, from 1945 to 1951, the BBC was the principal voice of the West and the leader in hours of international broadcasting. As other nations expanded their radio effort, the BBC, with its finances restricted, lost its lead in terms of broadcast hours. But in terms of the quality and credibility of its programs, its reputation has been maintained.

Part of the BBC's success stemmed from its pattern of organization. The BBC External Services are an integral part of the BBC and operate under the same Royal Charter, and the same director-general, who is the corporation's chief executive officer. He heads the permanent staff, acting under a board of 12 governors, each appointed for a five-year term. It is this charter and board of governors that give the BBC External Service its independence from the Foreign Office, oblige it to reflect British public opinion as a whole, and afford a freedom of action in broadcasting without official restriction or censorship. It is this organizational pattern that enables, for example, the British ambassador in Moscow, in response to a Soviet complaint about broadcasts on dissidents, to state that the Foreign Office does not control these broadcasts.

On the other hand, the BBC gives a full presentation of official British foreign policy by means of official statements and interviews, and this BBC role as spokesman is made clear in the broadcasts. The Foreign and Commonwealth Office has primary authority in determining what languages will be used, and the Treasury, through the Foreign and Commonwealth Office grants, has control over the grants to BBC. The Foreign Office must keep the BBC informed about foreign policy, and may query individual broadcasts ex post facto to ensure that no sustained orientation against government foreign policy arises in programming. Broadcasts are generally "centrist" in terms of British politics, but editorial control clearly remains with the BBC.

The BBC's External Service is able to draw on the news resources of the domestic radio and television services in its coverage of world events, and—particularly in times of crisis—its broadcasts are widely relied upon as a source of accurate and dependable news. From the trials and experience of the 1930s

and the 1940s, the BBC has developed procedures and performance standards that, in 1973, produced over 300,000 letters from
listeners worldwide, including over 3,000 letters a year from Poland, 1,500 from Hungary, and over 1,000 annually from Yugoslav listeners (BBC, 1974: 59, 63). In 1975, the BBC External
Service was broadcasting 727 program hours weekly, in some 39
languages and dialects. The Eastern European language services
include Bulgarian, Czech, German, Hungarian, Polish, Romanian,
Russian, Serbo-Croatian, Slovak, and Slovene.

The budgetary trend in the BBC External Service worldwide
has been slightly downward over recent years, though less so for
the Eastern European services, reflecting a certain priority given
by the British Foreign Office to this area. Most transmissions to
Eastern Europe are from Skelton in England, where the transmitters have seen 25 years of service. Other transmissions are
from a Cyprus installation, now being modernized, from a South-
east Asian transmitter beamed at Central Asia, and from a transmitter in Berlin.

Something of a first cousin to BBC External Services is Deutsche Welle (DW), established in 1953. West Germany's worldwide
broadcaster, however, has more independence from Foreign Office control than the BBC, partly because it is funded by the In-
terior Ministry. It is, however, also more closely tied than the
BBC to party political organizations, owing to the composition
and method of appointment of its council. Under a law passed
in 1960, Deutsche Welle is supervised by the Broadcast Council,
composed of two members chosen by the Bundestag, two by the
Bundesrat, four by the Federal Government, and three members,
each chosen by one of the three major religious faiths—Catholic,
Protestant, and Jewish. The council lays down guidelines on program policy and elects the general manager on the recommenda-
tion of an administrative council. This body, selected by the
Broadcast Council, supervises the general manager's operations.

More often than the BBC External Service, Deutsche Welle is
in conflict with its Foreign Office. A particular case was the
broadcasting of Alexander Solzhenitsyn's *The Gulag Archipelago*.
DW considered that the book, a much-discussed best-seller in the
Federal Republic and thus an aspect of the German contemporary
scene, was a legitimate choice for programming. The Foreign

Office felt that DW had stretched its charter but could not interfere. German diplomats in Moscow could legitimately have responded to Soviet criticism of the broadcasts by pointing to the fact that the Broadcast Council, not the Foreign Office, laid down the governing guidelines.

Deutsche Welle, with over 90 programs in 34 languages, is worldwide in orientation, while another network, Deutschlandfunk, functions principally as a home service with a limited European service to East Germans, Czechs, Hungarians, Romanians, Serbo-Croats, and Polish audiences. An important technical difference is that Deutschlandfunk uses long and medium-wave, while Deutsche Welle use short-wave transmissions. DW is by far the largest of the two corporations, which together were on the air 806 hours weekly in 1973, in 38 languages—an expansion of almost 800 percent since 1955 (BBC, 1975: 70-71).

Among other West European international broadcasters, Radio Sweden transmits 21 hours a week in Russian (World, 1976: 126). Kol Israel broadcasts to the Soviet Union and Eastern Europe and, more extensively, Vatican Radio (World, 1976: 142, 174). Belgium and the Netherlands beam no broadcasts there, but Radio Luxembourg, a remarkably successful medium-wave commercial broadcaster, is aimed at both Western and Eastern Europe (World, 1976: 114), and its outstanding programs of music and entertainment draw a wide audience.

Radio Canada International

While there had been recommendations for an international broadcasting service both by the Canadian Broadcasting Corporation's (CBC) Board of Governors and by parliamentary committees since the 1930s, Radio Canada International (RCI) was not inaugurated until the end of February 1945, commencing its overseas broadcasting with a Czech service (RCI, 1973; Brown, 1976). Its number of language programs grew steadily—Magyar was added after the Hungarian revolt in 1956—until a task force in 1973 recommended stopping some West European services and increasing concentration on the East. It now broadcasts 150 hours a week in 11 languages, seven of which are East European: Czech, German, Magyar, Soovak, Polish, Russian, and Ukrainian.

It presents balanced programming of news, commentary, interviews, cultural activities, sports, and music (RCI, 1976; Brown, 1976). Programs are sent by cable to Daventry in England, Sines in Portugal, and to Berlin for relay on short-wave (RCI, 1976; World, 1976: 236).

Organizationally CBC, of which RCI is a part, is similar to the BBC, being governed by a board of 15 directors who are not political appointees but responsible only to the Canadian parliament. RCI reports to the executive vice-president of the board. While it also keeps in close touch with government departments in Ottawa, particularly the Department of External Affairs, the discussions are concerned not with program content, but with such subjects as determining areas of the world to which broadcasting is important for the Canadian national interest. There is no government supervision or censorship, and in its programming RCI enjoys even more freedom of action than the BBC External Service (Brown, 1976).

The Voice of America

After 1945, both the public and Congress felt a strong distaste for an official American peacetime information organization, an attitude nourished by fears that it could become a domestic political force and the belief that private media were the appropriate instruments for public information. VOA's programming was almost halved, and its future seemed highly uncertain. The crises of the late 1940s, however, brought a marked reversal of congressional attitudes. The 1948 Smith-Mundt Act gave VOA legislative sanction, funding it as part of the State Department's program of international information and educational exchange. The early leaders of this effort stressed the value of straight presentation. One of them, Edward Barrett (1953: 6-7), expressed it thus:

> Totalitarian tyrants are miles ahead of us in recognizing the growing force of mass opinion. Stalin painstakingly built an international propaganda mechanism, from training schools up, that now consumes far more than a billion dollars a year. Our job should be easier and less expensive. We don't have to traffic in falsehood and distortion. If we can wield the truth effectively enough, if we can back our high aims with equally high actions, and if we can tell of them persuasively, we will be well on our way.

It was Barrett who persuaded President Truman to launch the major "Campaign of Truth" in April 1950 in similar terms:

> We must make ourselves heard round the world in a great campaign of truth. This task is not separate and distinct from other elements of our foreign policy. It is a necessary part of all we are doing [Sorenson, 1968: 26).

The Korean war stimulated a vigorous expansion of U.S. foreign information efforts. Funds were tripled, the role of overseas communication and its importance for foreign policy gained increased (though often reluctant) recognition and status, and the program was given more autonomy.

The Voice of America operates differently from the BBC or DW. VOA is much more what its title says—the official voice of the executive branch, which conducts American foreign policy. VOA, an integral part of the United States Information Agency (USIA), is charged with three tasks: to serve as a reliable, objective source of news, to present U.S. policy, and to portray American society. The Secretary of State is required to furnish policy guidance, but it is given administrative independence by being under USIA, an independent agency reporting to the president.

From the outset, there have been clashing concepts of the role of VOA, as well as a national ambivalence about overseas information activities generally. For a time in the early 1950s, VOA was a favorite target of Senator Joseph McCarthy's vicious and demoralizing attacks. Since then, as the most conspicuous and costly element in U.S. overseas communication programs, VOA has often been made the target of domestic debate and controversy. Despite a history of vicissitudes, VOA has developed major professional and technical skills in international communication.

Today the largest number of VOA broadcast hours (168 weekly) are beamed to the Soviet Union in Russian, Ukrainian, Estonian, Latvian, Lithuanian, Armenian, Georgian, and Uzbek. Other East European countries receive 87 hours. VOA has by far the largest audience of any Western broadcaster to the Soviet Union. Soviet listeners turn to VOA for its explanation of U.S. policy, its portrayal of the American lifestyle, and its music, especially jazz.

There has been debate in the 1970s about how VOA can increase its political effectiveness and, especially in the light of the communications revolution, reach more listeners. At the heart of the debate is the fact that VOA, unlike the BBC and DW, has direct ties to government policy. It does not have a semi-public board, such as the BBC Board of Governors, to buffer it from some sensitivities of diplomacy, and therefore is less free to broadcast forthrightly, for example, on human rights issues or on certain news developments. In times of crisis, especially, the latter have posed problems. During the first hours of the Soviet invasion of Czechoslovakia in 1968, VOA broadcast only straight news on the subject. Its audience naturally turned to BBC, RFE, or elsewhere. More recently, VOA has been criticized for not broadcasting more news about Soviet dissidents and about such best-sellers in the West as *The Gulag Archipelago*. Despite these omissions, VOA probably leads all broadcasters in the topicality of its news, certainly as related to the policies or events emanating from Washington.

There are clearly two opposing schools of thought in the debate on VOA's organization. The first holds that VOA must retain its official status—either kept as an integral part of a government overseas information agency, or else incorporated into the State Department itself; in any event, carefully circumscribed by the State Department. Otherwise, the argument goes, it will be little more than a commercial international broadcaster, and congressional support for it would wane. The other view, held notably by Senators Jackson and Percy, is that VOA needs the latitude made possible by the BBC format so that it can speak out more forthrightly and effectively (Stanton, 1975).

THE SURROGATE HOME SERVICES

Radio in the American Sector, Berlin

A unique foster-child of USIA, but quite separate from VOA, is the Radio in the American Sector, Berlin (RIAS). Despite the division of occupied Berlin in 1945 into four sectors, a proposed agreement with the Soviet Union stipulated that all four occupying powers would share in the control of Radio Berlin, the most powerful and important station in Germany. Characteristic of the anomalies of the Berlin situation was the fact that Radio Berlin was located in the British Sector, but since the Russians had occupied it first, it remained something of a Soviet oasis.

The American authorities tried hard to get Soviet agreement to Allied participation in Radio Berlin, and the first plans for RIAS were designed more as pressure on the Soviet occupiers than as a project to develop in earnest a station of our own. Only when the Soviet authorities remained adamant in their refusal, and our military government's need for an outlet became imperative, was the Drahtfunk inaugurated on February 7, 1946, working through a hook-up with the telephone system. The Soviets, through their control of the German postal administration (which included the telephone network), tried to sabotage the early efforts to establish RIAS. This was an ominous portent, since officially relations in the winter of 1945 were still quite good. When all hope of shared control of Radio Berlin had to be given up, RIAS acquired increasingly powerful transmitters. It steadily expanded its broadcasting role, through medium and short-wave frequencies and frequency-modulation transmissions, to reach all East Germany's population of 17 million, as well as East Berlin.

The role of RIAS in maintaining the morale of the Berlin population during the Berlin blockade and subsequent other crises was of critical importance. The station has a special political role in symbolizing the U.S. government's commitment to West Berlin. At the same time, unlike any of the other international radios broadcasting into communist territories, it became a joint German and American venture, receiving financial support as well as a commitment of facilities from the Bonn government. The transmitters have always remained under U.S. ownership, but all employees except the director are German. In another regard, too, RIAS is unique. With most transmitters located in Berlin and surrounded by East Germany, it is not readily subject to jamming.

Opinion surveys among visitors from East Germany and East Berlin indicate that RIAS is considered the most popular station among 80 percent of those questioned, and it is estimated that three-quarters of East Germany's adult population are listeners to RIAS.

Radio Free Europe and Radio Liberty

Radio Free Europe (RFE) and Radio Liberty (RL) differ markedly in origin from VOA and indeed from other Western broadcasters; to a very large degree they differ in programming and functions as well. VOA's primary responsibility is to present and explain U.S. institutions, culture, society, and official policies to a worldwide audience. RFE—broadcasting to Poland, Czechoslovakia, Hungary, Romania, and Bulgaria—and RL—broadcasting to the peoples of the Soviet Union—devote the bulk of their programming to developments within those countries or matters of direct concern to them, especially events about which news is often suppressed, distorted, or neglected in their own censored media.

Because RFE and RL together broadcast more program hours to the Soviet Union and Eastern Europe than any other foreign broadcaster, in more languages, and because of the significant evolution that has taken place in their mission and program concepts, their history and present unique roles have particular significance for a communications diplomacy that might foster a constructive East-West relationship.

Both radios began at the height of the Cold War. RFE had its beginnings in 1949 when the then "National Committee for a Free Europe" (incorporated in New York that year; now Free Europe, Inc.) sought ways to enable exiles from Eastern Europe to broadcast to their own countries. Such a radio had the backing of George Kennan and his Policy Planning Staff. In 1950 RFE went on the air with a mobile 7.5-kilowatt short-wave transmitter in West Germany. By the year's end it was broadcasting to Romania, Poland, Czechoslovakia, Hungary, Bulgaria, and (for a brief period) Albania. The following year, RFE had three transmitters in Germany and one in Portugal (operating under a Portuguese-licensed, American-funded corporation with the acronym RARET). Programs were prepared in Munich and New York.

RL began as "Radio Liberation" under the auspices of the "American Committee for the Freedom of the Peoples of the U.S.S.R.," incorporated in Delaware in 1951. The committee became the present Radio Liberty Committee in 1959, and Radio Liberation was later renamed Radio Liberty. Broadcasts did not begin until 1953, just a few days before Stalin's death, from two 10-kilowatt transmitters located in Lampertheim, West Germany.

RFE and RL were established in a political climate of escalating tensions, when the policy of "liberation" was still discussed as a viable American option, when people talked of "roll-back" and "anti-communist crusades." At the outset, RFE considered that its mission was "to sustain the morale of the captive people and stimulate in them a spirit of non-cooperation." It sought to "remind listeners that they were governed by agents of a foreign power," to display "the moral and spiritual emptiness of communism," and to inculcate "hopes of eventual liberation" (Library of Congress, 1972b: 22). RL in its first broadcasts took a similar approach to its audiences: the initial guidelines spoke of "implacable struggle against communist dictatorship until its complete destruction" (Library of Congress, 1972a: 9).

This roll-back and liberation emphasis was soon revised. After the death of Stalin, developments within the Soviet Union and Eastern Europe, and changes in United States foreign policy brought changes in programming. "Liberalization" replaced "liberation;" rhetoric was more restrained; polemic increasingly gave place to straightforward coverage of events, both within the countries to which programs were directed, and on the world scene. With increased broadcasting experience, standards of reporting became increasingly professional.

This evolution, sustained and accelerating throughout the 1960s, produced a far-reaching transformation in RFE's and RL's view of their purposes, and in the style and content of broadcasts. In 1972 RFE's internal policy guidelines stated:

> RFE's role is to provide a wide spectrum of facts, analyses and opinions in order that the peoples of East Europe be able to form their views on the basis of maximum relevant information.

The guidelines warned against vituperative, vindictive, or belligerent material, as well as programming that could be construed as unrealistic or inflammatory. Broadcasts must not "in any way lead the East European peoples to believe that in the event of an uprising . . . the West would intervene militarily" (Report, 1973: 83, 87).

Today RL notes in its guidelines that the more brutal the facts, the less emotional should be the presentation, and its commentators must avoid polemic with Soviet media. "Radio Liberty's role is that of an independent radio devoted to the dissemination of objective, balanced information." It approaches the world scene and internal events "from the point of view of peoples within the U.S.S.R. who share an interest in free speech and free access to information as a basis for a greater and more effective participation in the decisions affecting their lives and the country's place in the world community," providing ideas and information to audiences "as a basis for formulating their own ideas and finding their own solutions" (Report, 1973: 88-89).

In their genesis in the years following World War II, when the Soviet Union was imposing its hegemony in Eastern Europe, it was considered that RFE and RL programs would have greater credibility if they appeared to be privately funded, rather than supported by U.S. government funds. Hence financing of the radios—RL entirely, and RFE over 90 percent, with the balance supplied by public fund-raising campaigns—was provided by Central Intelligence Agency (CIA) funds. As this arrangement became known in the late 1960s, there was strong public and congressional criticism of such covert support. In 1971 Senator Clifford Case publicly questioned this mode of financing and introduced a bill for open government funding. Senator J. William Fulbright and other critics, viewing the radios as "relics of the cold war," and seeing them as still in the mold of the 1950s, questioned whether it was in the national interest to continue their operations, particularly by government funds, in an era of détente.

Between 1971 and 1973 the structure and operations of RFE and RL were the subject of intensive studies—by the Congressional Research Service of the Library of Congress (1972a, 1972b), by the U.S. Comptroller General (Comp. Gen. of U.S., 1972), and by a special Presidential Commission on International Broadcasting under the chairmanship of Dr. Milton Eisenhower (Report, 1973).

All these studies suggested various improvements, but uniformly commended the performance of RFE and RL in filling the information gap created by government control of information in the Soviet Union and Eastern Europe. In accordance with the recommendations of the Eisenhower Commission report, "the Right to Know," a Board for International Broadcasting (BIB) was established (U.S. Cong., 1973: 2) as an independent public body to oversee the operation of the two radios. Five public members were to be nominated by the president and confirmed by the senate, with membership to be drawn from those experienced in international affairs or telecommunications. The functions of the new board include making grants from appropriated funds to the radio corporations, reviewing their mission and operation, assessing the quality, effectiveness and professional integrity of their broadcasts, and developing and applying methods of evaluation to ensure that the grants "are applied in a manner not inconsistent with the broad foreign policy objectives of the U.S. Government." BIB reports annually to the president and the congress, with the secretary of state providing information on the foreign policy of the United States, rather than "policy guidance," as in the case of VOA. In case of disagreements between the State Department and the radio corporations on the substance of broadcasts, BIB is empowered by law to be the arbiter.

The Act creating BIB, passed by broad bipartisan majorities in both houses of Congress and signed into law in October 1973, declared that RFE and RL "have demonstrated their effectiveness in furthering the open communication of information and ideas" in Eastern Europe and the Soviet Union. In the following year, $49.8 million was appropriated for the radios, arousing little of the controversy that had marked earlier years. Despite recession, in 1975 the congress showed extraordinary support for the radios by a 30 percent increase in the budget to allow RFE and RL to be consolidated and some transmitting facilities to be modernized.

This increased funding enabled the purchase of 12 transmitters (100 KW each) for modernizing the short-wave transmitting stations at Biblis and Lampertheim, Germany. One medium-wave transmitter is at Holzkirchen, near Munich, where most programming is done. In need of modernization is the major RFE facility at Goria, Portugal, which has 18 transmitters, some dating back to the 1950s. This facility has always been organized under a Portuguese corporation with joint Portuguese and American (RFE) directors; this joint-venture approach helped the installation weather the storm throughout the recent crises in Portugal. Comparatively less familiar to the Spanish public is the Radio Liberty facility, with no such joint company, at Pals, Costa Brava, Spain. From six transmitters (five 250 KW, and one 100 KW) at this ideal location by the Mediterranean, on the second ionosphere hop, Radio Liberty can reach Central Asia. As in Portugal, the lease agreement for this facility is due for renewal; this facility is also in need of modernization.

The two radios had evolved a new role and different programming well before the link with CIA was ended. The transformation carried with it some difficulties. The staffs of both radios included many emigrés, often recent arrivals, from the Soviet Union and Eastern European countries. They had unrivaled knowledge of the broadcasting styles and patterns of their home countries, in addition to their language skills; and some were well-known and highly respected among their audiences. (The most effective of them, reports show, are considered by their listeners not as foreigners and outsiders, but as free voices of a domestic operation.) For some, however, rejection of communist regimes, reinforced by bitter experiences, occasionally led to a highly partisan and polemic stand. To meet this problem, both RFE and RL have increasingly insisted on a careful distinction between reporting and commentary. They keep a careful watch to ensure that broadcasts do not become instruments of any particular emigré group or position, but that all responsible points of view are objectively presented to listeners. In rare moments of candor the Soviets themselves recognized the change. As early as 1967, Artem Panfilov concluded:

> In practice, propaganda for the overthrow of the "communist regime" has almost disappeared from all American broadcasts to the socialist countries of Europe. Even "Radio Free Europe" no longer broadcasts such propaganda. . . . The tone of the radio broadcasts has changed significantly: the crude insinuations and profanities have disappeared. Direct interference in the internal affairs of one country or another in the form of all sorts of advice to radio listeners has almost ceased, and undercover propaganda has left the scene [Library of Congress, 1972a: 16].

The radios are strongly news-oriented, with news forming almost a third of their programming. Both news and features, aimed as an alternate "home service" to national audiences, stress developments in their countries, in neighboring communist countries, and East-West relations. For RFE and RL to speak in their audiences' terms requires that programs be especially well-informed and up-to-date on internal developments and circumstances within those countries. They have developed impressive research services to ensure this knowledge, and the scholarly quality of their studies has led to their use by all West European foreign offices as well as by the BBC and DW.

Radio Liberty audience surveys, validated by independent scholars such as Professors Ithiel de Sola Pool of MIT and Seymour Martin Lipset of Stanford University, indicate that in the course of a month the service reaches between 35 and 40 million Soviet citizens. RFE audience and public opinion research, validated by Oliver Quayle, indicates a listening East Eruopean audience of about half the adult population. Both RL and RFE base their findings largely on interviews with Soviet bloc citizens travelling in Western Europe; RFE, drawing on a more accessible sample of East Europeans, has Austrian, Danish, French, and British independent polling institutes conduct the interviews. The following table and graphs indicate the broadcast hours of each of the 25 languages in which RL and RFE broadcast.

TABLE 1

Radio Liberty Programming	Original Programming Per Day		Total Hours Broadcast	
	Hours	Minutes	Per Day	Per Week
1. Russian	5	19	24	168
2. Byelorussian		50	6	42
3. Ukrainian	1	5	15	105
4. Estonian		30	4	28
5. Latvian		39	4	28
6. Lithuanian		39	7	49
7. Armenian		49	6	42
8. Azerbaijani		37	6	42
9. Georgian		46	6	42
North Caucasian:				
10. Avar		6	2	14
11. Chechen-Ingush		6	2	14
12. Ossetic		6	2	14
13. Tatar-Bashkir		5	6	42
Turkestani 1:				
14. Uighur		10	1.83	13
15. Uzbek		26	2	14
Turkestani 2:				
16. Kirghiz		12	.83	5.8
17. Tadzhik		12	.83	5.8
18. Turkmen		12	.83	5.8
Turkestani 3:				
19. Kazakh		25	1.7	11.6
Radio Free Europe Programming				
20. Bulgarian	6		8.13	57
21. Czech/				
22. Slovak	9		20.5	144
23. Hungarian	10		18.7	131
24. Polish	10		19	134
25. Romanian	7	11	13	91

Table based on BIB programming data, 24 November, 11 December 1957. (Reprinted by permission of the Board for International Broadcasting, Washington, D.C.)

**RADIO LIBERTY PROGRAMMING
BY SUBJECT CATEGORIES,
FISCAL YEAR 1975**

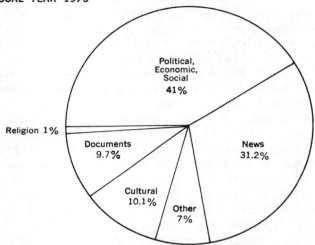

**RADIO FREE EUROPE
PROGRAMMING BY
SUBJECT CATEGORIES,
FISCAL YEAR 1975**

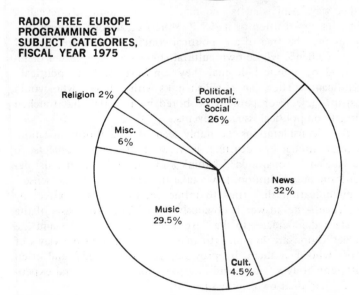

Graphs from BIB Second Annual Report (1975): 45
(Reprinted by permission of the Board for International Broadcasting,
Washington, D.C.)

The Nature of Propaganda

We have been discussing international broadcasting largely in terms of its political role—that is, directed to foreign audiences whose listening motives are primarily political interest or curiosity. We should recognize another dimension—cultural programming—broadcasts that present or deal with music, the arts, the humanities. Programs of this kind are presumably directed toward those motivated by cultural interest or curiosity. Some, perhaps, merely seek entertainment.

But the lines are not easily drawn. On the one hand, an audience that seeks out and listens to the music of a foreign broadcaster is manifesting a community of interests, a sharing of common pleasures and human creativity. This is no trivial bond. On the other hand, when a nation offers illustrations of its own cultural life to a foreign audience (whether in broadcasts or in cultural exchanges, exhibitions, or overseas libraries) it is presumably doing more than merely asking for admiration. It is expressing something significant about itself that can be said in no other way. And what is said may be more eloquently revealing than its constitution or its GNP. Moreover, though cultural programs may be free of any political content, if they are directed to an audience whose own cultural choices are suppressed or restricted by political dogma, they can have implicit "political" significance. They can establish links with listeners who would distrust, reject, or perhaps be bored by material that touched directly on political issues or events.

Such considerations inevitably lead us to a thorny question. Is international broadcasting, in view of its political aspects, to be labeled "propaganda?" I think a meaningful answer must depend on our definition. If we take the term in its broad sense— communication that tries to influence, that aims to affect its audiences—the answer is obviously "yes." The broadcasts of the Western democracies to the Eastern European countries and the Soviet Union are intended to make a difference in the views of those who hear them, by providing access to a wider and often different body of facts and interpretations—and cultural experience—than their own sources permit.

But if we take the word in its pejorative implications—implications it carries for many Westerners—as signifying sinister manipulation, deceit, and concealed motives, the answer must be "no." The broadcasts of the Western democracies do not seek to impose some contrived and self-serving version of the world and dictate assent by rigging the evidence. That was the Goebbels model of propaganda, resting on a cynically concocted mixture of falsehood and truth. It was essentially psychological coercion, a national weapon deployed on the terrain of mind and emotion to force a response.

The broadcasts of the West, certainly, are selective: no medium can provide *all* the facts, all the possible interpretations. They speak as the voices of the Western democracies, reflecting inevitably the biases of their historical experience and their values. To that extent they are not objective, and lay no claim to clinical detachment. The BBC, despite all its devotion to impartiality, makes this point explicit:

> Impartiality does not imply an Olympian neutrality or detachment from those basic moral and constitutional beliefs on which the nation's life is founded. The BBC does not feel obliged for example to appear neutral as between truth and untruth, justice and injustice, freedom and slavery, compassion and cruelty, tolerance and intolerance (including racial intolerance). This is an important reservation, but not one which detracts from the BBC's overall determination to be impartial in its presentation of controversial issues [BBC Handbook, 1974: 281].

But Western broadcasters do not distort their messages to impose some predetermined reaction or some monolithic orthodoxy, choosing rather to reflect their own societies' diversities and divisions. They accept competition in the marketplace of ideas. They are committed to telling the truth; if peaceful and enduring international order is achieved, it must be based on fact. And they are committed to an open world, where the free flow of information and ideas makes dialogue possible.

These broadcasts are intended to affect their audiences, but as communication, not manipulation. To the extent that they can restore to their hearers the human right to a free flow of information and ideas, they move toward dialogue. And, as Jacques Ellul has memorably observed, "Propaganda ceases where simple dialogue begins" (1965: 6).

III. SOVIET CONTROL OF COMMUNICATIONS

Communication and its control have been fundamental concerns of Soviet leadership from the beginning of the Soviet state.

Perhaps the salient characteristic of Lenin's version of Marxism, especially of his conception of the disciplined hierarchical revolutionary organization out of which grew the Bolshevik party, the Soviet state, and the international communist movement, was his emphasis on the power of political communication [Barghoorn, 1964: 5].

Communication had been Lenin's route to power, achieved by his vigorous pamphleteering and notably by his emigré newspaper, *Iskra* (The Spark). Agitation and propaganda and the ruthless defense of ideological orthodoxy are part of the genetic inheritance of the Soviet regime.

These concerns are also rooted in Russia's history. The autocratic rule of the Tsars had included uneven but often rigid censorship, excluding and repressing ideas that might challenge Tsarist authority. But Soviet control over the flow of information, ideas and people—among its own citizens, and from the outside world—is the most elaborate and pervasive system ever known. Under it, all the channels of public communication serve to define, disseminate, and maintain ideological cohesion and a public, official version of events and issues. The Communist Party of the Soviet Union, through its control of all Soviet society, has attempted almost complete indoctrination and censorship.

Soviet methods of control go far beyond ordinary censorship. They become a positive injunction, a more subtle and pervasive coercion. A persecuted writer, the late Arkady Belinkov, described censorship as practised in the Soviet Union as "a new phenomenon in the history of thought control." Previous dictatorial societies had been concerned "merely to repress heretical opinions, whereas the Soviet Communist Party has introduced a system so thorough that it not only censors a writer but dictates what he shall say" (Dewhirst and Farrell, 1973: 1). In this system, he notes, the censor has played a secondary role. The totalitarian system aims to shape the entire process of literary, artistic, and journalistic production. The writer "is surrounded and conditioned by the ideology that has spawned this censorship" (Dewhirst and Farrell, 1973: 3). The pressures to conform are omnipresent and unremitting, and the evidence of the system's coercive power inescapable. Censorship thus becomes a continuous process, a seamless web of direct and indirect influence and compulsion. Editors and writers, by anticipating constraints, thus become involuntary collaborators in the censorship process. There arises a kind of self-censorship, an internal thought-police in the individual mind that subjugates will and imagination.

Even for members of the Soviet elite, who have access to special sources of information relatively free of general censorship, information is subjected to screening. The bulletins they receive vary with their rank. Western observers often wonder how reliable is the information prepared for the Soviet leadership; even for them it is elaborately compartmentalized and hierarchically distributed.

Various causes have deepened Soviet concern with communication. In the post-Stalin era, the role of political communication has been extended as the Soviet Union moved to great-power status. Thus Leonid F. Ilichev, deputy minister of foreign affairs, stated at a Moscow conference in 1960 that the Party had now made "ideological work and the communist education of the working peoples" the central task, and that the success of communism as a movement depended largely upon the abilities of the communists as communicators (Barghoorn, 1964: 7). Soviet leaders saw opportunities in new regions, among new nations, and in new conflicts and uncertainties of interests. With the need for increased ideological range and adaptability, increased vigilance and control were required to ensure that permissible limits were strictly set and observed, and central doctrines not eroded.

Soviet concern with communication has been further intensified by de-Stalinization and the accompanying "thaw," the selective abatement in secret police terror, the publicized policy of coexistence, and expanded contacts with the West. These raised the likelihood of Soviet citizens questioning the leadership's credibility and spurred new calls for vigilance. Increased exposure to the West was to be kept within careful limits and its effects vigilantly counteracted. In 1969, the editors of *Sovietskaya Rossiya*—the party newspaper of the Russian Soviet Federated Socialist Republic, the largest Soviet republic—wrote that tourism, economic relations, and cultural exchange were all exploited by imperialists to "prettify capitalism and defame socialism," to arouse discontent in the communist countries, and to undermine faith in the "ideology of socialism." The editors argued that the party therefore made special demands on those working on the ideological front—writers, journalists, people in the arts—as the "vanguard of the struggle against the troubadours of imperialism" (Hannah, 1972: 19). The party, said the editors, "has entrusted to them the most powerful weapons—the press, radio, television, the theatre and cinema."

Soviet controls can and often do operate with great efficiency. The mass media are maintained as monopolies by party and state. Direct and indirect censorship assure that content accords with centrally-determined policy. Nothing may legally be printed without the censor's approval—from state decrees to the labels of vodka bottles. Domestic radio, cinema, stage, music, works of art—all must receive official approval. Organizations like the Union of Soviet Journalists and Union of Soviet Writers assist the censorship apparatus by serving as vehicles for disciplining professional activities and enforcing conformity. Acceptability determines access to the media, career advancement, and rewards—or punishments. Even tighter controls try to ensure that nothing from abroad will give the Soviet citizen facts or ideas that might conflict with official Soviet versions, or could stimulate questions about orthodox interpretations of the past, the present, and the future.

The Soviet assertion of a policy of peaceful coexistence and desire for increased access to the West entailed a certain modification of Stalin's rigid and hostile exclusionism. "Normalization" of relations, official postures of amity, and great power "respectability" required some authentication. In his radio-television address to the American people during his 1973 visit, General Secretary Brezhnev said:

> To live at peace we must trust each other, and to trust each other we must know each other better. We for our part want Americans to visualize our way of life and our way of thinking as completely and correctly as possible.

But opportunities for the Soviet people to know Americans and the West have been severely circumscribed. Exchanges of exhibitions and cultural performances are subject to strict Soviet government agreement and control. Item by item reciprocity is generally imposed, and the number, content, and itineraries are carefully limited and laid down.

The recent flow of hundreds of thousands of western tourists has inevitably provided Soviet citizens with some opportunity to encounter foreigners. The KGB, despite its enormous apparatus, cannot follow and watch closely every westerner travelling in the Soviet Union. Often foreign tourists sit at the same table with Russians in Soviet restaurants. Inevitably, conversation follows. But here the invisible barrier of thought control prevents meaningful communication. Many Soviet citizens suspect, and sometimes with good reason, that contacts with travellers from the West may mean trouble, and they can display a remarkable ability to shut out foreigners without actually offending them. Traditional Russian cultural isolation and defensiveness toward strangers are often additional restraints.

Certainly, since Stalin's day, the regime has allowed more Soviet citizens to travel abroad. However, candidates for travel have to go through a complicated selection process designed to eliminate those who do not have reputations of uncompromising political loyalty to the party. Yet these travelers gain new perspectives on Soviet life from their experiences outside Soviet borders. This author has frequently been impressed by the thoughtful objectivity and frankness of some Soviet visitors. But their behavior back home, in the Soviet Union, is completely different, and their independent-mindedness disappears. For example, Soviet reporters covering the United States seldom convey to their readers their real impressions of life in the United States. Most of their stories could just as well be written in the editorial offices of *Pravda* and *Izvestiya* by somebody who had never visited America.

A fairly large number of Western publications are received in the Soviet Union, but they are made available in a highly selective way only to special groups. Special permission is needed if a Soviet citizen wishes to subscribe to any noncommunist foreign publication, and such permission is granted only to very senior officials. Even newspapers published by foreign communist parties are occasionally confiscated if they carry stories contrary to official Soviet policies. Issues of the French communist newspaper *L'Humanité* criticizing the treatment of dissenters or other Soviet policies have never reached their Soviet readers. Scientific and technical literature, particularly, is assiduously collected for official use by the Soviet scientific establishment, but its distribution and availability in libraries, laboratories, and institutes is very selectively controlled. A specialist will have available just that material needed for his work but will have great difficulty in obtaining wider access.

For the general public, the Soviet criterion for access is even more arbitrarily selective. Arthur Hailey's *Airport* was allowed to become a best-seller in the Soviet Union, but Bibles cannot be imported. The period of détente, it is true, has seen increased importation of officially-sponsored foreign publications such as USIA's magazine *America*. This is permitted a monthly circulation of 62,000 and the British publication *Anglia* has a quarterly circulation of 104,000. Both are covered by reciprocal agreements, but the reciprocity is something of a sham since the sale of Soviet publications in this country or in Great Britain does not require official agreement. Moreover, the Soviet Union does not comply strictly with the agreements. *America* and *Anglia* are not available on newsstands in many cities where they are supposed to be. Although top officials and foreign policy analysts receive these publications at home, they can currently be purchased only at selected newsstands in a few large cities by those willing to come early in the morning on the day of delivery and wait in long lines, or by those who have special arrangements with the vendor.

The Problem of Foreign Broadcasts

But in one area of communications, effective controls are difficult to maintain: radio broadcasts from abroad. Many Soviet citizens heard of Stalin's death from the controlled wired loudspeaker, used at the time as the prevalent means of broadcast communication with the populace at large. Wired-radio diffusion to the masses continues, largely in places of public gathering, such as factories and hotels. But since that day in 1953 there has been a technological revolution in radio broadcasting and in television. The Soviet regime has demonstrated that it appreciates the new potential both for furthering mass political indoctrination and for foreign penetration of information.

In the latter half of the 1950s, the number of radio receivers in the Soviet Union more than tripled: over 20 million were in use by 1960. By 1964, wave radio receivers outnumbered wired speakers, rising to over 35 million. In 1974 there were, by official count, 57.1 million short-wave sets, but this figure was again surpassed by the number of wired sets, up to 59 million, which are accessible to most Soviet citizens, especially in European Russia.

Within the Soviet Union, AM broadcasting is by medium and long-wave for metropolitan areas. Short-wave is used to span the vast reaches of the largest country in the world. Because of this dependence on short-wave communication with its own people, the Soviet regime cannot ban access to short-wave sets. It has, however, tried to limit their capabilities; although most sets can receive short-wave, bands are often quite restricted, so as to eliminate frequencies best suited for reception of foreign broadcasts. But such measures are often defeated by the fact that electronics training is widespread in Soviet schools and youth organizations, and the scarcity of repair facilities has encouraged many people to become their own radio technicians. In this way, many have learned to modify sets so that they can receive foreign broadcasts more readily (Hannah, 1972: 115).

Foreign short-wave broadcasts thus remain a major point of vulnerability in communication control. The Soviet Union has sought to meet this problem in a variety of ways. Because of its own use of short-wave broadcasting to reach the Soviet people, it has never made listening to foreign broadcasts a criminal offense per se; however, distribution of hostile information and propaganda is considered an offense, and sometimes includes the mere repetition of material from foreign broadcasts. The tremendous growth of radio listening since the mid-1950s, moreover, has made official prohibition impractical.

One method of combating the influence of foreign broadcasts has been the attempt to discredit the content of programs, and the motives, sometimes even the personalities, of foreign broadcasters. Especially in times of tension, the use of such methods has been stepped up in Soviet press and radio. For example, in the Czechoslovak crisis year of 1968, *Izvestiya* articles linked the BBC and Britain's Secret Intelligence Service (SIS).

> The BBC, the mouthpiece of naked anti-communism in the British Isles, has been taking a very active part in several extremely ugly SIS operations. This is confirmed by a number of documents in our possession. . . . In connection with all this, another document has come to light; it states that "an agreement exists with the BBC to turn over to the SIS all letters sent to the BBC from listeners in the socialist countries" [Hannah, 1972: 115].

This latter allegation was clearly a hint to the Soviet citizen that letter-writing to the BBC might entail a charge of cooperating with British intelligence. Soviet authorities are aware that attacks on foreign broadcasts, and direct attempts at rebuttal and refutation can also serve to advertise foreign programs and their awkward content. It was at one time the practice to refrain from attacking Radio Liberty (Hannah, 1972: 114-115).

Another response has been an effort to make Soviet domestic broadcasts more competitive, particularly by making their news content more comprehensive and topical, as well as by increasing their liveliness and variety, often in direct imitation of Western programming practices. In the mid-1960s surveys of listeners were instituted, giving some indications of the preferences, size, and structure of audiences. Publication of results has been fragmentary, and without better knowledge of the data base and methodology, evaluations cannot be conclusive. We have little information about how Soviet authorities have interpreted these surveys. A recent study by a Western scholar summarizes the results as follows:

> About 40 to 60 million people, with varying degrees of regularity, listen to foreign radio broadcasts. Major questions of public interest that are known mainly through foreign radio coverage reach and are of interest to 50 to 75 percent of various population groups. From 30 to 50 percent of the population consider the response of Soviet broadcasting to be inadequate. In addition, from 20 to 30 percent of the population, and perhaps more, seem generally to doubt the credibility of all Soviet information sources, and by inference, much of the basic ideological legitimacy of the system [Lisann, 1975: 155].

Disquieting though such figures might be to Soviet officials, no major revolution in Soviet radio has been forthcoming. Central direction, bureaucratic caution, and insistence that radio is essentially an ideological instrument have worked to keep news stale and selective, and to inhibit efforts to lighten program dullness.

There have been, however, major changes in format which, while functioning in the same basic way, have "packaged" the message better in order to compete with western broadcasts. One sophisticated counter to the rising audience for foreign broadcasts was the initiation in July 1964 of a new Soviet domestic program, *Mayak* (Beacon). It offered a streamlined, more attractive blend of commentary, popular music, and frequent news programs. And it broadcast around the clock, sometimes on the frequencies used by foreign broadcasts, thus also serving as a jamming vehicle.

Jamming

Jamming, of course, has been a direct and conspicuous weapon against the entry of uncensored information via the air waves.

Orginally, jamming consisted of broadcasting noise, usually on the same frequency as the unwanted program together with Morse identification signals. Either of two propagation techniques were employed: long-range via sky-waves reflected from the ionosphere, or by local ground-wave. In sky-wave jamming, the jamming transmitter must be located roughly as far from the receiving area as the broadcaster. Ground-wave jamming, on the other hand, used around urban centers, is very short-range and requires the use of clusters of many transmitters to cover large population centers.

Jamming is seldom completely effective, but when it is intense, it can drive away listeners, especially the less committed, and can greatly decrease the attractiveness of entertainment and music. Even in the areas of most intense interference, such as Moscow and Leningrad, some signals normally get through both sky-wave and ground-wave jamming. Reception can be improved by the determined listener who is willing to experiment—for example, by moving his radio set around to find a spot where clearer reception is possible. Many city-dwellers, furthermore, go to the country on weekends, where reception is generally good. Tape recordings of the most interesting programs taken from international broadcasting are circulated among friends.

There is a propagation phenomenon known as "twilight immunity," when the signal can be received with little or no interference on short-wave. During late afternoon in the target area, the propagation path from the West is still in daylight. A sky-wave jammer station, in order to achieve the requisite distance, may be located far to the East where all is in darkness. Because propagation conditions vary widely with light and dark, a given frequency may be inoperative from the jamming site. The twilight-immunity phenomenon, however, does not apply to ground-wave jamming for large cities, since the interfering signal operates over a very limited distance and does not vary in strength with the time of day.

The Soviet policy on jamming has varied in relation to East-West tensions, but more particularly in consonance with crises, internal or external. The Soviet Union first began jamming in 1948, directing it against the Russian-language broadcasts of the Voice of America initiated the previous year. In 1949 the BBC was jammed; soon the satellite countries joined in, and jamming was extended to RCI, Deutsche Welle, RFE, and Radio Liberation, now Radio Liberty.

The first relaxation came during six months of 1956, with Krushchev's visit to Britain, when only RL and RFE were jammed.

Poland stopped interference with RFE in November 1956, resuming in 1971. The Hungarian and Suez crises brought renewed jamming until 1960, when Prime Minister Harold Macmillan successfully induced a few months' respite. This was aborted by the U-2 crisis. From July 1963, when final negotiation of the Test Ban Treaty began, until August 1968, the Warsaw Pact countries suspended most jamming; Romania ceased to hamper RFE transmissions in the summer of 1963 and Hungary stopped in March 1964. Bulgaria and Czechoslovakia have never ceased to jam RFE. Kol Israel was jammed on the outbreak of the 1967 Middle East war. The invasion of Czechoslovakia in 1968 brought about an increase of jamming.

The next relaxation came in September 1973, the eve of the opening of negotiations on the Conference on European Security and Cooperation at Geneva. The Soviet Union ceased jamming most western broadcasts while continuing interference with transmissions from those of the People's Republic of China, Albania, Israel, and RL (Hale, 1975: 138). In Eastern Europe, jamming of RFE continued for the Czechoslovak, Bulgarian, and (intermittently) Polish services. Partial action against VOA by Bulgaria was finally ended in 1975. Selective jamming of Kol Israel has been maintained, varying with Soviet interest in discouraging Jewish emigration (Hale, 1975: 138). Soviet jamming of RL has been uninterrupted.

The Soviet Union continues interference against some stations despite its inefficiency and high cost. In 1971, it was estimated that the Soviet Union spent about $300 million a year on jamming—six times the cost of its international broadcasts (Report, 1973: 70-71). When the Poles ceased jamming for a period beginning in 1956, they themselves gave the amount saved as $17.5 million, which was then the equivalent of the worldwide VOA budget (Hale, 1975: 132). Recent estimates of the number of jamming transmitters in the Soviet Union run as high as 3,000 (Report, 1973: 70; Hale, 1975: 133). This willingness to lavish resources on a somewhat ineffective means of keeping out uncensored communication is eloquent testimony to the penetration of international broadcasting and to the desire of citizens in the Soviet Union and Eastern Europe for freer access to ideas and information.

Jamming is contrary to international law. The Soviet Union has signed the Montreux International Telecommunications Convention, which provides that:

> All stations, whatever their purpose, must be established and operated in such a manner as not to cause harmful interference to radio services or communications of other Members or Associate Members.

Jamming is also a violation of Article 19 of the Universal Declaration of Human Rights, which states that:

> Everyone has the right of freedom of opinions and expression; this right includes freedom to hold opinions without interference, to seek, receive, and impart information and ideas *through any media and without regard to frontiers* [emphasis added].

As dissenters have noted, jamming is also a violation of the 1936 "Stalin constitution."

In 1950, the Sub-Commission on Freedom of Information and the Press, established by the UN Commission on Human Rights, studied the issue of jamming and presented the Economic and Social Council with a draft resolution, which was submitted in revised form to the General Assembly. Opponents argued that countries had the same right to defend themselves against hostile propaganda as against opium smuggling or the entry of pornographic literature. In the Soviet view, unless a state is permitted to bar the inflow of news and opinions considered subversive, the free flow of information is an infringement of sovereignty. The great majority of countries argued, however, that states should raise no such barriers to freedom of information and that the listener had the right to judge the value of the program. The resolulution was passed on December 14, 1950, by 49 votes to 5. While urging all governments to refrain from radio broadcasts that represented "unfair attacks or slanders against other peoples anywhere," the resolution noted that some countries "are deliberately interfering with the reception by the peoples of those countries of certain radio signals originating beyond their territories" and it condemned "measures of this nature as the denial fof the rights of all persons to be fully informed" (Codding, 1959: 75).

The Soviet model for communications has been adopted in most communist countries but there have been some notable exceptions to Leninist authoritarianism on this issue. For example, when the Poles in 1956 gave up jamming for a period, they acknowledged that it was "a method that has brought us no credit," and that stopping it "is a victory for the principle, correct in our opinion, that foreign radio stations should be countered by factual argument" (Hale, 1975: 139). If this principle were firmly adhered to, and the "rights of all persons to be fully informed" respected, international radio and the Soviet responses to it could become an important vehicle for the kind of rational dialogue needed for constructive international relationships.

The Problem of Television

Technologically, television from abroad may become another challenge to the Soviet information monopoly in the 1980s, when satellite TV may enable signals to be beamed directly into domestic sets. Soviet television sets numbered around five million in 1960; today there are probably about 60 million sets in operation, and the number is increasing, making it possible for almost any Soviet citizen to see television, except for inhabitants of sparsely settled areas in Siberia. Anticipating this potential vulnerability, the Soviet Union in 1972 presented a draft resolution in the United Nations General Assembly that direct television broadcasts should be declared illegal without the consent of the receiving country. The Soviet aim was to have all satellite TV beamed without such consent branded as interference in internal affairs, and to obtain legal international sanction for countermeasures.

The United States opposed what became known as the "Jammers' Charter" and took a position supporting the free flow of information and ideas. The Soviet Union moved the issue to the UN Committee on Peaceful Uses of Outer Space, where the issue could be debated in the context of cooperation in this field (Hale, 1975: 135-136).

The position of the Soviet Union in this matter is paradoxical. It objects to information coming into the Soviet Union, endeavoring to prevent the possible employment of television for that purpose and using all possible means to obstruct foreign radio broadcasting; at the same time it is becoming the largest purveyor of broadcast information to the world (Report, 1973: 14-20; BBC Handbook, 1976: 65).

IV. THE SOVIET UNION: FERMENT, REPRESSION, AND CHANGE

E arlier we noted the role of censorship, indoctrination, and propaganda as pervasive forces in Soviet society that not only censor the writer but seek to dictate what is thought or written before pen is put to paper. Has the Soviet system succeeded in creating a mechanism of conditioning and control, of coercion and insulation, imposing near complete conformity and intellectual obedience? Has the Soviet Union become like George Orwell's fictional society, where the "thought police" and the "Ministry of Truth" not only suppress any intellectual opposition but also transform human beings into computer-like, programmed, easily-manipulated creatures?

About 20 years before the climactic Orwellian date of 1984, this issue was taken up by an eminent authority on political communication, Ithiel de Sola Pool. He concluded on the evidence of massive research (de Sola Pool, 1965), that the full realization of 1984 had been prevented, largely because of international communication:

> Most of the things of a positive character that are happening in the Soviet Union today are explainable only in terms of the influence of the West, for which the most important single channel is radio. There is now enough communication to keep us influencing each other, to assure that any Western idea circulates in the Soviet Union, too. The pessimistic expectation that totalitarianism could develop an accepted heinous civilization of its own by 1984 or any other year has been defeated primarily by the forces of communication, and above all by international radio.

But what role had indigenous ideas and ideals played in aborting a strange and utterly alien culture, and in keeping East and West in some recognizable degree a single civilization? For, although ideas of democracy and liberalism have—in the long run—never successfully challenged the dominance of despotism in Russia's history, movements of protest and reform, of dissent and revolt, have also marked that turbulent history. These have stemmed not only from the East-West interface that has always played an important part in Russian life, from the "Westernization" imposed by Peter I and Catherine II, and the ideas of the Enlightenment that helped to stimulate the revolutions of the eighteenth and nineteenth centuries. They have also sprung from native roots—the deep humanity, the passionate concern with justice, the anguished self-scrutiny that find their eloquent expression in the great works of Russian literature.

The Temporary Thaw and Reaction

What role had the heritage of such internally-generated ideas and values in the political and social process—in Eastern Europe as well as the Soviet Union—in forestalling an Orwellian "1984"? In 1965, when Professor de Sola Pool made his somewhat sweeping statement, the answer to this question was that the active influence of indigenous ideas was minimal or marginal: they were effectively repressed. But there was one major exception: for a brief but seminal period, there was a "thaw."

In the late 1950s the initial shock of de-Stalinization and its attendant revelations had been given added force by the personal accounts of terror and cruelty many heard from friends and relatives released from labor camps and political prisons in the "thaw." The literary world, in Russia the traditional area of nonconformist political discussion and resistance, took up these themes and some publication of novels and stories dealing with them was permitted.

Alexander Tvardovsky, the editor of the country's most important literary magazine, *Novy Mir,* took advantage of this opportunity by calling on Soviet writers to tell the full truth about Stalinist tyranny. Khrushchev personally allowed the publication of Solzhenitsyn's prison camp novel, *One Day in the Life of Ivan*

Denisovich, first serialized in *Novy Mir* and widely read. A cheap edition of 750,000 copies published early in 1963 was quickly exhausted. The book was also avidly read in Eastern Europe. Zhores Medvedev was to write that the publication of this novel "it seemed, could only help to consolidate the democratic process. It encouraged hope that in its wake other writings, polemical, historical, or scientific, would find their way into print . . . works that would not attempt to ignore the experiences of the past." This book, he said, "was therefore seen not only as marking a new stage in the development of Soviet literature but also as a sign of improvement in all aspects of intellectual life in our society" (Medvedev, 1973: 4).

But such hopes were ill-founded. In 1964 Khrushchev was deposed; the KGB was ordered to recommence literary harassment; censorship was tightened; more self-censorship was demanded; and Stalin was partially rehabilitated. Solzhenitsyn's second novel, *The First Circle,* was denied publication and the other works Medvedev had hoped for were not to be printed, at least not in communist countries.

Thus the literary ferment was well under way when the new Soviet leadership determined that this new liberalism was dangerous and should be snuffed out. But the authorities were to be disappointed. It was the suppression itself that was responsible for producing a whole new genre of Soviet literature in the form of *samizdat* (self-publishing)—mimeographed and typed manuscripts passed from hand to hand. It became a cottage industry, transmitting a new intellectual and cultural life within Russia.

But at best such underground industry could reach only thousands of Soviet citizens. International broadcasting rescued the new literature from this limitation and found for itself an entirely new dimension of programming. The principal impact of international radio prior to 1964, Professor de Sola Pool noted, was to spread Western ideas beyond the iron curtain and to preserve something of the humane traditions of liberal Western culture. But the new suppressions beginning in 1964, combined with the new genre of *samizdat* literature, gave a further role to international broadcasting—the dissemination to millions within the Soviet bloc what *samizdat* was communicating only to thousands.

The 1966 trial of Andrei Sinyavsky and Yuli Daniel detonated a chain of events in the dissent movement that was to drive it into the open. Sinyavsky, who now lectures at the Sorbonne, was one of the best-known Soviet literary critics. Daniel was a gifted young literary translator. Both men were highly regarded in literary circles. What the regime did not know was that the two men were also underground novelists, published under pen names in several Western countries. Eventually, however, the KGB discovered this and in the early autumn of 1965 Sinyavsky and Daniel were arrested. Their subsequent trial was a kind of propagandistic happening designed to intimidate literary dissenters. They were charged with anti-Soviet propaganda, and sentenced to long terms in hard labor camps. It was the first case in Soviet history where artists were officially sentenced for their literary work.

Their conviction spurred broader and bolder protest. A self-styled "Democratic Movement" emerged with regular publication of a *samizdat* news bulletin, "Chronicle of Current Events." *Samizdat,* at first a chiefly literary and artistic effort, took on a reportorial function, chronicling abuse and acting as an opposition press. The Soviet government took strong countermeasures. The KGB stepped up house searches, arrests, trials, and misuse of mental hospitals. But many dissidents expressed defiance in open, signed protests, and worldwide publicity resulted.

The Dissident Movement

The dissident movement encompasses various schools of thought. The "Democratic Movement," includes the eminent nuclear physicist, Andrei Sakharov, who helped to form the Committee on Human Rights to protest Soviet practice as violating Soviet law. Sakharov, while conceding that Russia may not be immediately ready for Western-style democracy, asserts that democracy must be the ultimate goal.

There is also a school of Marxist reformers, nondemocratic in that they subscribe to the basic principles of Leninist centralism. To this school belong the Medvedev brothers, Roy, a historian, and Zhores, a biochemist, twins who consider themselves loyal Marxists. They have protested the tyranny of Stalin and its mass murders, and they have also attacked the more sophisticated methods of the present regime which puts dissenters into mental hospitals. (Zhores Medvedev's experiences during incarceration in a mental hospital became the basis of his book, *A Question of Madness*.) Despite intense criticism of the Soviet system, this "loyal opposition" believes that the hope of reform is from within and that this process is not helped by outside pressures such as the Jackson·Amendment, which would make external détente conditional upon internal reform.

Another group, whose best-known member is Alexander Solzhenitsyn, like the nineteenth century Slavophiles look to the traditions and values of "old Russia"—including the "purity" of the *Veche,* a form of medieval popular assembly. They fear the stress of industrialization and economic growth, call for environmental concern, and seek, not a Western-style democracy, but a benevolent autocracy rooted in the values and beliefs of the Russian Orthodox Church. Their views find expression in the *samizdat* publication, *Veche.* Solzhenitsyn disagrees with the Medvedev brothers' belief that external pressures will hinder internal reform. Like Sakharov, he believes that détente must be linked to reform from within and concessions from the regime.

The Nationalities Problem

An entirely different aspect of the dissident movement has its roots in the ethnic composition of the Soviet Union. Out of 240 million inhabitants, just over half are Russians. The other Slavs—Ukrainians and Byelorussians—constitute developed nations with their own distinct histories, traditions, cultures, and languages. Georgia, Armenia, and Azerbaijan enjoyed independence for centuries. Estonia, Latvia, and Lithuania were forcibly incorporated into the Soviet Union in 1940. Other ethnic groups of Turko-Tatar, Finno-Ugrian, Caucasian, and Mongolian origins continue to cultivate their heritages. They and *not* the Russians may comprise the majority of the Soviet population, if the present birthrates continue for a few more years.

Throughout the world, more than 140 nations—many·of them never before independent—have now been recognized by the world community. They are able to live, work, and advance their national interests as fully sovereign countries. The non-Russian peoples of the Soviet Union know about this—in no small measure through international broadcasting. There is ample evidence that many of these people consider themselves to be victims of the world's last surviving colonial empire.

Increasingly since 1972, many nationalist activists have been arrested. In the Baltic areas, particularly in Catholic Lithuania, religious unrest has continued for years; not long ago more than 47,000 people signed petitions demanding religious instruction for their children. In Vilnius, the Lithuanian capital, there were public disturbances. Muslim nationalities, too, notably the Crimean Tatars, have produced their dissenters and resisted Russification. The degree and forms of nationalist feeling among the minority peoples in the Soviet Union of course vary considerably, and often differ in their emphasis on either full national independence or, with self-government, remaining in the Soviet federal structure.

The Jews of Russia present a special case that has attracted world attention. The Jews rank among the 15 largest nationality groups, with a population of between two and a half and three million; unlike most other nationalities, they are territorially dispersed. There has been anti-Semitism in Russia for centuries. The largest entry of Jewry into Russian history, which followed the eighteenth century partitions of Poland, met with harsh discrimination. Pogroms and harassment of Jews in nineteenth century Russia aroused resentment in the United States. In 1869, President Ulysses Grant protested to the Russian government the proposed expulsion of 20,000 Jews from southwest Russia. In 1892 the House of Representatives refused to allocate funds for food transport to Russia in protest against Russian treatment of Jews.

With the fall of the Tsar in 1917, discriminatory legislation was repealed; this enlightened policy was continued by Lenin, with prominent Jews such as Trotsky and Zinoviev in the party leadership. Under Stalin, discrimination returned; his daughter Svetlana writes of the dictator's anti-Semitism increasing with age. After a brief and partial relaxation accompanying de-Stalinization, it became evident that anti-Semitism would persist in work, education, and travel. The response of Soviet Jews was the emigration movement of the late 1960s and 1970s.

As emigration increased, dissident and *samizdat* links with the Western world strengthened. Correspondents of the Western press increasingly focused their human-interest reporting on dissidents, repressions, and discrimination. The Soviet population could learn of these from foreign broadcasting of the reports of these correspondents.

Soviet concern over the emigration issue has been heightened by the nationality problem. Although intellectual dissent is more vocal and attracts more immediate attention, over the long term the nationality dissent is probably far more worrisome for Moscow in hard geopolitical terms. The authorities were confirmed in their fear that other groups would grasp at the precedent established for Jews. Although many nationality movements (like the Jews at first) claimed cultural autonomy within the Soviet Union, some changed their goals on seeing the success of Jewish determination and solidarity. Under the Soviet law allowing emigration for the purpose of reunification with families abroad, Volga Germans now applied for emigration to West German families, and Meshketians hoped for "reunion" with families who had emigrated to Turkey centuries ago.

The Issues of Détente and Stalinism

An issue dividing Soviet intellectual dissidents is the relationship of détente abroad to reform within. In an August 21, 1973 press conference, Sakharov stated:

> Détente without democratization, a rapprochement when the West in fact accepts our rules of the game . . . would be very dangerous . . . and wouldn't solve any of the world's problems, and would mean simply a capitulation to our real or exaggerated strength. It would mean an attempt to trade, to get from us gas and oil, neglecting all other aspects of the problem. . . . By liberating ourselves from problems we can't solve ourselves, we could concentrate on accumulating strength, and as a result, the whole world would be disarmed and facing our uncontrollable bureaucratic apparatus. . . . That would be . . . encouragement of closed countries, where everything that happens goes unseen by foreign eyes behind a mask that hides its real face. No one should dream of having such a neighbor, and especially if this neighbor is armed to the teeth [Tökés, 1975: 386].

The debate has been echoed in the West by Senator Jackson and Secretary Kissinger, the latter arguing that détente helped make such dissent possible. While it is true that Soviet efforts to promote Western cooperation in trade and other matters have restrained Kremlin policies, compared to Stalin, it is also true that such restraint has not applied to more sophisticated forms of tyranny.

Public appeals by well-known dissidents became increasingly common practice. On September 14, 1973, Sakharov wrote an open letter to the U.S. Congress endorsing the Jackson Amendment. On January 5, 1974, Sakharov, the poet Alexander Galich, the novelist Vladimir Maximov, and others signed a declaration of concern over the new threats to Solzhenitsyn contained in a TASS statement. TASS claimed that the Nobel laureate was a "traitor to the Motherland," libelling its past. How was it possible, the signatories asked, to claim that the "mistakes made" had been corrected and then to characterize as libel Solzhenitsyn's attempt to publish historical information? "After all," they argued,

> it cannot be denied that there actually were mass arrests, tortures, executions, forced labor, inhuman conditions, and the deliberate annihilation of millions of people in camps. There was the dispossession of the Kulaks, the persecution and annihilation of hundreds of thousands of believers, forcible resettling of peoples, anti-worker and anti-peasant laws, and the persecution of those who had returned from prisoner-of-war camps [Salisbury, 1974: 232-233].

The issue which united all dissident schools of thought was the truth about Stalinism. Roy Medvedev, a communist party member, spent six years in the 1960s preparing a massive exposé of Stalin entitled *Let History Judge* (1971). Publication in the Soviet Union was denied, and it was eventually circulated through *samizdat*. Medvedev was unrelenting in his criticism of those who had previously concealed the truth about Stalin. The Soviet despot, he wrote, would "never be forgiven his monstrous crimes. Neither will those historians, political leaders, and writers who lacked the courage to study the history of Stalin's crimes be forgiven" (Medvedev, 1971: 274).

Western scholars added to the weight of the evidence and the charges. Adam Ulam, in his biography of Stalin, asserts that the Soviet Union is still Stalin's Russia:

It is only when the Soviet people are able to look at their recent past and recognize it for what it really was—tragic and heroic, certainly, but also and in many ways preposterous—that the spell will be lifted and the Stalin era will finally have ended [Ulam, 1973: 741].

Whether the "Stalin era" did or did not end is an important question because of its international implications. Dictatorial control in the manner of Stalin, particularly in the flow of information, can nullify international agreements, through reinterpretation, and even the failure to report that they exist. Resurgent *samizdat* literature, by revealing the truth of Stalinism, can "lift the spell" once it reaches Soviet and Eastern European peoples. International broadcasting can help it to reach them.

The Audiences of the Future

Many present day listeners to foreign broadcasts are either dissidents or thoroughly disillusioned with the regime and its brand of communism. But many other groups, still basically committed to the present power structure, are beginning to feel doubt. Estimates of the number of Soviet listeners for all foreign radio stations range as high as 60 million (Lisann, 1975: 128). Yet the potential audiences are far greater.

From the Soviet government's own research, we know that the majority of Soviet citizens are basically dissatisfied with the quality of information and news that they receive. We know, too, of their thirst for accurate information on foreign affairs. In fact, the Soviet readers have a high interest in the very areas where they are mostly denied accurate, up-to-date, and complete information. Major Soviet newspapers have polled readers to determine their preferred topics (Hannah, 1972; 62-73). The *Izvestiya* study reported:

Topics	Percent of Readers Mentioning Topic
Conflict situations	87
Foreign affairs	47
Feuilletons (satirical articles or sketches)	46
Work of the Soviets	39
Accidents and crime	33
Economics	5

A later report on *Izvestiya's* readers placed "propaganda articles" and "work of the Soviets" at the bottom of listener interest. In a third survey of readers of *Trud,* "events abroad" attracted 74 percent; and "articles on international themes" drew 63 percent. "Party themes" drew 40 percent of readers, and "official communications" was the topic of least interest at 18 percent. Dr. Hannah (1972: 74) makes a cogent observation regarding the desire of the Soviet citizen to have a freer flow of information than the regime is willing to countenance:

> One would assume that the regime would like people to read articles on economic progress and political matters, and certainly the amount of space devoted to these topics in Soviet newspapers would indicate just that. The audience surveys, however, leave us with the impression that the most popular rubrics are international news, family circle topics and news about accidents and crime (to which Soviet newspapers devote very little space).

Clearly the newspapers fail to win readers for regime propaganda, their primary mission, while their rigid censorship deprives their readership of accurate foreign news as well as internal human interest stories.

Modernization of Soviet society is further expanding audiences and their needs. Since the late 1920s, the country's socio-occupational structure has altered greatly. In 1928 nearly 80 percent of the population were peasants; by 1970 the proportion of the agricultural labor force had declined to less than 30 percent (Katz, 1974).

Modernization has also meant mass migration from the countryside to the cities, with the children and grandchildren of peasants now firmly entrenched in an urban lifestyle. In little more than half a century, the Soviet urban population has increased from 22 million to more than 150 million. By 1970, the year of the last census, 10 Soviet cities had a population of over a million, while a dozen more were approaching that size (Katz, 1974).

Even more dramatic has been the increase in intellectual workers; from two million in the Russia of 1913 to an estimated 35 million today. This movement has been accompanied by increased entry into professional careers, especially among the younger and middle generations; professional consciousness has been further intensified by the mushrooming of professional organizations of specialists.

These movements have paralleled a massive spread of education among the Soviet population. The number of graduates and enrollees of institutions of higher and specialized secondary education has been doubling every dozen years or so. The regime's stated objective is to give every citizen at least a full secondary education, and in larger cities this campaign has neared complete success. These developments have brought added importance, and increasingly attentive listeners to international broadcasting—especially Radio Liberty, in its function as a horizontal communications flow, or cross-reporting, among people at local levels in the various republics. The Soviet media are so hierarchical, so centrally organized, that events occurring in one region of the Soviet Union (such as the bread riots in Rostov in the early 1960s) are unknown to people living in other areas. This prevents the development of a sense of common problems at the grassroots level.

The Soviet communications network might be likened to a hub with spokes extending in all directions to the republics and localities, but with no wheel rim to connect the spokes. For example, a Ukrainian resettled in Kazakhstan cannot subscribe to the hometown newspaper back in the Ukraine. Only the national republic, or local papers for a given administrative region are available. Local information reaches other regions mainly through word-of-mouth and foreign broadcasting. International broadcasting here facilitates something other than the vertical, or center-periphery, information flow controlled by the party through its news services and media.

The essential drama of the Soviet future is the sharpening contrast between the population's rising levels of education and aspiration, and the dreary stagnation of economic and political development. Increasing numbers of the Soviet elite are now aware that the Soviet Union, of all major industrial countries, has by far the lowest living standard, and that in terms of political progress their country lags behind many far more primitive societies. Bitter jokes are frequently heard contrasting the Soviet Union's foreign aid or its space achievements with its continuing shortcomings on the home front.

Even in a cohesive, homogeneous society tightly knit by a common national identity, such contrasts would evoke a demand for more objective standards and information than official sources provide. In the Soviet case, these tendencies are increased by the centrifugal pressures of non-Russian nationalism.

Economic managers, top scientists and experts, some elements of government, and occasionally even the party bureaucracy and certainly the youth, have indicated growing interest in Western ideas and culture. An intelligent, sophisticated contribution to discussion among members of the elite becomes an increasingly important task for international broadcasting. Recent arrivals from the Soviet Union report that people who hold important positions, who are not outspoken dissidents, are more and more distrustful of the official media and seek other sources of information. They are the people who will help determine the future of Russia and its international conduct.

V. INTERNATIONAL BROADCASTING TO EASTERN EUROPE

The combined population of the six Warsaw Pact countries and Yugoslavia, some 130 million, is greater than that of France, West Germany, and Belgium together. Each country has a character of its own, shaped over the centuries by shifting tides of migration and settlement, successive invasions and conquests, and geography that in some regions has facilitated conquest and invasion, in others presented a barrier that defended and isolated. They have known the imperialisms of East and West, the rivalries of Eastern Orthodoxy, Western Catholicism, and Turkish Islam. Though they have produced rich cultures and even empires of their own, the peoples of the area have seldom controlled their own destinies for long; powerful neighbors have generally determined their borders and sought to dictate their allegiances.

As peoples, they are very old; as nations, they are very recent; as governments, their communist regimes were all brought to power by Soviet domination and support after World War II. In this area are to be found a great range of ethnic diversity, languages, and folkways. As nations they differ in internal cohesion and homogeneity, in the level and direction of economic development, in the rigidity of their internal controls, and in their degree of independence from Soviet policies. Today a traveler to Eastern Europe is struck by the fact that, more than three decades after coming under Soviet control, they continued to differ strikingly from the Soviet Union and from each other, a visible sign of the strength of surviving nationalist feeling and influences in the face of increasing pressure for Soviet-dictated communist conformity.

In one fundamental regard the situation in these countries is different from that in the Soviet Union. The latter is a super-power with nuclear weapons, a well-developed industrial economy, a pioneer in space exploration, dominating most of Eastern Europe and seeking to expand its power to other, non-European areas of the world. Many Soviet citizens may keenly feel their lack of freedom, and their relatively low living standards; but for most there is also a strong sense of national loyalty and pride in their nation's power and accomplishments. The Soviet regime responsible for these results is not seen as an alien presence imposed from without; however much Russians may be conscious of the heavy hand of state and party, or unhappy with the demands of dogma, they link party and government leadership with national achievement, and the regime is at great pains to cultivate this patriotic identification. Among the nationalities who have been subsumed in the Soviet Union, of course, the success of this identification is strongly qualified, and often resented as a form of "Great Russian" chauvinism.

But to the peoples of Eastern Europe, Soviet-Russian dominance is felt to be imposed by political maneuver and military power; the regimes that are seen as its instruments enjoy no such patriotic identification or legitimacy. The citizens of these countries seek not only greater freedom internally, but greater freedom from Soviet—and historically Russian—dominance as well. Ties and affinities with Western nations are often deeply rooted in their culture and history; they respond to Western ideas not only as an alternative to communist domination, but as a natural part of their heritage. Thus they provide responsive audiences to Western broadcasts.

If Eastern Europe is generally more affluent than the country that dominates it, that is due to a combination of factors: some nations, like Czechoslovakia, had a much higher economic and educational level in their precommunist days; many trade extensively with the noncommunist world (ranging from 71 percent of all Yugoslav foreign trade, through 58 percent and 51 percent for Romania and Poland, down to a low of 27 percent for Bulgaria); and finally, it has been quite clear that the Soviet Union was prepared to make economic concessions of a kind denied to its own people in exchange for the political stability and docility of Eastern Europe.

The people of these countries seek to increase and strengthen their national independence. Much evidence of this can be found not only in the stand of Czechoslovakia's Alexander Dubcek, but among more conservative communists. They all realize, however, that in the face of Soviet power, full independence is not a practicable choice, as the Soviet "intervention" of 1968 in Czechoslovakia dramatized. Romania, which pursues a course of limited independence in foreign policy, is careful not to go too far in antagonizing Moscow. Even Yugoslavia, not a member of the Warsaw Pact, treads cautiously. In the Eastern European countries, therefore, the issue of internal freedom is directly tied to relations with the Soviet Union.

The Effect of International Broadcasting

The most dramatic example of the impact of international broadcasting is to be found in Eastern Europe. The Soviet system brought many of the same problems to Eastern European nations as it did to nationalities subsumed within the Soviet Union. The Stalinist concept of "nationalist in form; socialist in content" weakened their historic traditions, cultural values, and folkways. It particularly sought to undermine their sense of national identity, attempting to eliminate a major source of potential resistance to Soviet influence. By allowing the peoples of Eastern Europe uncensored access to facts about their own lives and those of their neighbors, international broadcasting has helped to preserve national heritages, defend legitimate national interests, and widen the sphere of intellectual-cultural freedom.

A striking example of the latter is East Germany which, though described as "the ideological fortress of Eastern Europe," is at the same time particularly open to Western broadcasting. The 17 million inhabitants of the German Democratic Republic (GDR) are improving materially, with the highest per capita gross national product (GNP) of any country in the Soviet bloc. Deutschlandfunk's home service to East Germany and the presence of RIAS in Berlin, surrounded by the GDR, alone guarantee that East Germany receives extensive Western radio; many of the national services broadcast there as well. Not only do they have a very wide audience, but at least 80 percent of East Germans receive West German television broadcasts on ordinary sets.

For the past five years the communist authorities have resigned themselves to this wide general viewing of television until there is now no stigma or prohibition attached, with the possible exception of discouraging school children. There have been no attempts to jam Western television. The East German author, Stefan Heyman writes that "the influence of Western television and radio on the minds of people here is considerable," that it shapes "tastes in fashion, music and film; it creates consumer demands that GDR industry and trade make belated efforts to fill" (Osnos, 1976).

The open reception of radio and television stands in marked contrast to the banning of books and periodicals from the West. The East Germans have emphasized in their press the problems of the West, such as crime and inflation, which often appear openly on television.

A delegate at the recent GDR party congress, Professor Karl Tschnik, observed in an interview "it is not enough to prohibit viewing. That is too primitive." He felt an improvement of East German material and cultural life was the effective response (Osnos, 1976). In a country already well-off by Soviet bloc standards, such an argument reflects the considerable influence of Western radio and television broadcasting.

Hungary has recovered economically from the Soviet military reprisals which it suffered in 1956. This has been due in great measure to the introduction, in recent years, of a more flexible economic policy which has raised some ideological eyebrows. Internally, it is the most relaxed of the countries with Soviet occupation forces. Its 10.6 million citizens have a distinct ethnic tradition that sets them apart from neighboring Slavs. Hungary's culture is still reminiscent of precommunist days; it is strengthened by traditions going back more than a thousand years and strongly colored by later centuries of Austrian-Hapsburg rule, until World War I secured its independence.

The Hungarian revolt and Western reaction in 1956 not only discredited the Dulles doctrine of liberation, it also involved international broadcasting in controversy. Both the BBC External Service and RFE reported developments throughout the crisis. Later, the Soviet Union, as well as some Western critics accused the RFE broadcasts of encouraging an uprising that had no chance of success. A replay of taped RFE broadcasts absolved the organization, but even more stringent broadcast guidelines were issued.

The Hungarian revolt, although brutally suppressed, was not entirely in vain. The postrevolutionary regime under Janos Kadar introduced numerous reforms, and today Hungary's independence is similar to that of Poland. Early in 1964, the Hungarian regime stopped jamming all Western broadcasting, which was perhaps another indication of its relatively moderate internal policies. RFE broadcasts to Hungary 18.7 hours daily, with an estimated audience of half the population; VOA broadcasts two and a half hours and the BBC two and three-quarters hours.

Poland consists of what remained after the Soviet Union had annexed a substantial part of eastern Poland and parts of Germany were added to the country. Poland now has one of the most ethnically homogeneous populations in the world (98.7 percent of its 34 million citizens are Poles) and the largest territory in Eastern Europe. It differs from its Warsaw Pact neighbors in at least two important respects: the continuing powerful influence of the Catholic Church despite the country's communist regime, and the fact that most Polish farmland has been restored to private use—a legacy of the "Polish October" brought about by the unrest of 1956. Its agriculture, and its rapidly expanded industrialization and diversification, are particularly vulnerable to Soviet pressures and tightened control.

In Poland, RFE is often called "Warsaw's fourth radio station," so great is its relevance to indigenous Polish concerns; Polish listeners are estimated at 56 percent of the population. Even the communist hierarchy listens and often forms its own editorial policy in reaction to the issues raised by RFE broadcasts. RFE transmits 19 hours daily to Poland, VOA two and a half hours, and the BBC three hours. The effectiveness of these broadcasts became clear in December 1970 when riots broke out as a result of the Warsaw government announcing increased prices of consumer goods, including food. The riots first occurred in the Polish Baltic seacoast cities, resulting in local police measures and censorship of news from the area. RFE, however, was monitoring local bulletins and so was able to relay to the entire Polish population news about the riots, the workers' demands, and repressive countermeasures.

General public reaction was strong. Party Secretary Wladyslaw Gomulka was replaced by Edward Gierek. Many workers' demands were accepted, repressive measures were relaxed, and living standards improved. Gierek was hailed by the Polish press as "more responsive to people's needs" than his predecessor.

A recent development, however, has possible ominous implications. Subsequent to the Helsinki declaration's affirmation of non-interference in internal affairs, a revision of Poland's constitution creates an obligation for Poland to retain its alliance with the Soviet Union, giving a new dimension to the Brezhnev doctrine.

That the Soviet Union did not intervene in Poland in 1970 was probably due not only to the strength of the feelings of the workers and uncertainty about the reaction of the Polish army, but also to the world-wide indignation caused by the invasion of Czechoslovakia two years before.

With a population of 15 million and continuing tough political controls, Czechoslovakia is—due partly to remnants of its prewar economic level—second only to the GDR in per capita GNP. In April 1968, the Czech government announced plans for a new constitution to include secret ballots and freedom of speech, assembly, and travel. Economic reform would cover market pricing, consumer industries, and decentralization of production and investment. Dubcek emphasized that the party would remain preeminent, industry would be publicly owned, and Czechoslovakia would remain a member of the Warsaw Pact. But he also wished to democratize communism, give it a "human face," and allow the free flow of information and ideas. These proposals were much more threatening than the primarily economic demands of the Polish workers. Their proliferation would menace Soviet control not only of Eastern Europe but also of its internal nationalities and intellectual life. Along with Warsaw Pact forces, the Soviet Union invaded and proclaimed the Brezhnev doctrine: that it has the right to intervene at will anywhere in Eastern Europe to maintain communism.

The fluctuation in foreign radio listening during this crisis is interesting. When the Stalinist-type leader Antonin Novotny was replaced by Dubcek, and the "Prague Spring" set in, with almost free media, RFE listening fell from 60 percent to about 37 percent of the adult population. With Soviet-led military intervention, the listenership shot up again. In 1969, Czechoslovak National Radio conducted an audience survey and ranked foreign radio broadcasts in the following order of popularity: RFE, Radio Vienna, BBC, VOA, and the German stations. A prominent Czech communist, Jan Kaspar, noted in the Party's monthly magazine: "The study proves the influence of the foreign stations is large and must not be underestimated." RFE broadcasts daily 20.5 hours, VOA two hours, and BBC three hours.

It has been foreign broadcasts that have kept alive the otherwise censored story of the events surrounding the "Prague Spring" and the Soviet intervention. When one of the leading figures of that crisis, Josef Smrkovsky, died in early 1974, RFE broadcast the letter written to his widow by Dubcek, who reaffirmed his beliefs, denounced the system, and stressed that Smrkovsky, contrary to official propaganda, had always remained a communist and had not opposed the Soviet Union, as Dubcek's successor, Gustav Husak, had charged. Dubcek called the present rule "a system of personal power from top to bottom," and asserted that the party had lost "what counts most"—the faith of the masses. Dubcek's letter was ignored by the official media. The Czechoslovaks learned about it from international broadcasting.

Bulgaria, although larger than Hungary, has the smallest population of the Warsaw Pact countries (just under 9 million). It is also thoroughly Slavic and the closest, culturally and politically, to the Soviet Union. This is due partly to residual Bulgarian gratitude to the Russians for having liberated them from Turkish rule in the latter part of the last century and, with one of the fastest growing economies in Eastern Europe, partly because of the vital flow of Soviet economic assistance, reportedly at the same level as Soviet aid to India despite the vast disparity in populations. A visitor arriving from the Soviet Union is struck by the living standard enjoyed by Bulgarians, in many ways better than that of their Soviet benefactors. RFE is the most popular broadcaster, with nearly half the population of Bulgaria as listeners (BIB, 1975: 24). It broadcasts over eight hours daily, VOA an hour and a half, and BBC two hours.

Romanians are proud of their ethnic and linguistic heritage, considering themselves unique in Eastern Europe as the descendants of Caesar's legionaries stationed in Dacia (the country's ancient name). Romanian remains a Romance Language. Despite its territorial losses (the Soviet annexation of the area now known as the Soviet Republic of Moldavia is a sore point in Soviet-Romanian relations), Romania is still large by Eastern European standards; it is nearly the size of neighboring Yugoslavia, and has more than 21 million people. There is a sizeable Hungarian minority (about eight million) mostly living in the Transylvanian area. Its determination to conduct a foreign policy in some respects independent from the Soviet Union means that it pays by having a lower standard of living than its neighbor, Bulgaria.

Among Warsaw Pact countries, Romania—with Nicolae Ceausescu's skillful cultivation of nationalism—has achieved the greatest degree of latitude in its foreign policy. In 1964, its central committee issued a declaration of independence from the COMECON, the focal organization in the Soviet-based economic integration of Eastern Europe. In 1967 Romania refused to break off relations with Israel over the June war, and in 1968 Ceausescu did not join in the invasion of Czechoslovakia.

Internally, however, Romania has remained second only to Bulgaria as a rigidly controlled communist dictatorship. There are signs of increasing ideological regulation of cultural life, such as the establishment in 1974 of "Editorial Central" to tighten censorship, and the "election" of a chairman to the Writers Union, who had, in 1971, disciplined the Union's weekly for opposition to the redogmatizing of cultural policy. Romanian spokesmen abroad argue that such rigid controls are a necessary part of the policy of independence from Moscow in foreign affairs, pointing out that internal events in Czechoslovakia led to the exercise of the Brezhnev doctrine. Another motivation is the desire to build up the industrial sector at the expense of the consumer sector in an effort to build national prestige, and to make Romania self-sustaining, relatively free of COMECON ties, and able to earn foreign exchange.

Nevertheless, international broadcasting faces a dilemma in commenting on programs and policies which simultaneously aim at close censorship internally and an independent foreign policy externally. An example of that dilemma was a most favored nation (MFN) trade agreement in 1975 with the United States, contingent upon more liberal emigration policies; but RFE, not the Romanian media, informed the Romanian people of this provision.

Jamming of all international broadcasting to Romania has ceased, probably in part because Bucharest rather than Moscow would have to foot the bill. It is noteworthy that the Romanian media attacks on RFE have been less vigorous than those of other Eastern European countries. RFE is on the air 13 hours a day to Romania, with 7 hours of original programming, and an estimated audience of 57 percent of the population. VOA broadcasts daily one and a half hours, and the BBC broadcasts two and one-fourth hours.

Yugoslavia is a very special case in Eastern Europe: although communist, under President Tito it declared its independence from Soviet control nearly 30 years ago. It has 21,300,000 people of heterogeneous ethnic and religious composition. It includes "five nationalities, four languages, three religions, and two alphabets" (Roberts, 1973: 5), which have given rise to six constituent republics and two autonomous provinces. The major split is between Serbia, a thoroughly Slavic culture whose language uses the Cyrillic alphabet and whose religion is Orthodox—and Croatia, which though Slavic, is a predominantly Catholic country using the Roman alphabet. With Western support it has moved toward a Western-type market economy.

Yugoslavia has achieved even more independence from Soviet domination than Romania, and also has a far more democratized communism, despite recent internal tightening. When RFE began transmissions to Soviet-dominated Eastern Europe, Yugoslavia had already broken with the Cominform and hence was omitted from RFE broadcasts. Yugoslavia is reached by VOA in both Serbo-Croat (one hour daily) and Slovenian (half hour daily) and also by the BBC (16 and a quarter hours weekly). Enough data are available to prove the great popularity of both VOA and BBC. For, even though the Yugoslav press and radio media report much more freely than those of other Eastern European countries, Yugoslavia, as a communist country, exercises important restrictions on freedom of information. Yugoslavs know this and therefore enjoy listening to uncensored foreign broadcasts.

Overall Impact of Broadcasts

On the whole, recent broadcasting to Eastern Europe has been primarily intended to supplement domestic stations and provide the population with alternative sources of information and analysis. A survey of potential listeners to RFE (almost 6,000 citizens visiting the West from six Eastern European countries) indicated that among the 51 percent who listened, the overwhelming majority (87 percent in Bulgaria and over 94 percent in all other countries) were either "very satisfied" or "more or less satisfied." Given that this is a tricky issue for travellers to discuss with strangers, a high degree of enthusiasm for RFE broadcasts is reflected. Positive response was weakest in the middle-age group.

In considering events in Eastern Europe, one important aspect is their secondary impact on the Soviet system, particularly when these effects can be enhanced by international broadcasting. Eastern Europe plays a significant intermediary role in public diplomacy between the West and the Soviet Union. The whole region acts as a permeable channel.

It has always been true that the movement of people has been greater among the "people's democracies" than the flow of tourism and cultural exchange between Western countries and the Soviet Union. Vacationers, cultural and educational figures, skilled workers—almost anyone travelling to Eastern Europe from the Soviet Union obtains greatly widened access to information and cultural materials not only from Eastern Europe but, more importantly, from Western Europe. All the people's democracies have more cultural contact with the West than does the Soviet Union. Western fashions, dances, musical and educational innovations—almost every kind of cultural output from 12-tone music to blue jeans has been passed through Eastern Europe on its way to the Soviet Union.

Some medium-wave broadcasts are regularly listened to by Poles and Hungarians: Vienna, Paris, Luxembourg, and others. Since there has been more travel between these Eastern European countries and the West than from the Soviet Union to the West, for most Soviet citizens contact has been possible only in Eastern Europe (including attendance at concerts and dance recitals, meeting foreign tourists, and so on). Some cultural events have been broadcast on Eastern European radio and television stations and heard by Soviet listeners.

All such contact between the Soviet Union and Western Europe through Eastern European countries might be termed "cultural seepage." It plays a very important role in informing Soviet citizens about the outside world, and much of its route is through foreign broadcasting to Eastern Europe and thence to the Soviet Union. This cannot but be helpful in maximizing the impact that broadcasting has on both, and possibly in bringing about a more humane system through pressures generated by responses to widened horizons and more extensive and sophisticated information among citizens in both areas.

VI. TOWARD A NEW COMMUNICATIONS DIPLOMACY

The opening chapter discussed the dichotomy in the Soviet Union's concepts and practice of détente. While Soviet leaders continue to proclaim that they seek a relaxation of tensions and broadened, more constructive relationships with other nations, they nevertheless increase their efforts to maintain the isolation, or at least the insulation, of the population. At the Helsinki conference the Soviet representatives appeared to make concessions to the free flow of people, information, and ideas. But their government promptly sought to impose qualifying interpretations on the document and to attack international broadcasting with renewed fervor.

From the viewpoint of Soviet leaders, the dichotomy is intrinsic to détente; indeed, it is fundamental to their entire conduct of foreign affairs. The duality in Soviet détente diplomacy is well under control and carefully managed. But it could be said that there is also a duality in American diplomacy, a duality not under control nor deliberately managed, and which sometimes seems to result in inner conflict and cross-purposes.

Democracies conduct their relationships with foreign countries through two channels. One channel is statist diplomacy: the formal intercourse between governments, through the meetings and correspondence of heads of state, through foreign offices and diplomatic representatives, through international bodies and conferences. Old as the nation-state itself, the conventions and apparatus of this form of diplomacy have developed over the centuries to respond to the mounting complexities of foreign relationships. But it is essentially a dialogue between the voices of official authority.

The other channel of relationships between nations is an outgrowth of democratic government, of the concept of popular sovereignty applied to foreign affairs. It conducts no negotiations, dispatches no notes, signs no treaties, presents no démarches. It has come to be called "people-to-people diplomacy"—the direct reaching out of peoples to speak to other peoples, quite apart from the formal operations of their governments. In the history of our own democracy there runs the strong belief that peoples at large, not simply their governments and rulers, are the legitimate concern of international relationships. In its most striking aspect, it is the conviction that we have an obligation to extend to the rest of the world the democratic freedoms we have achieved. Perhaps its fullest expression is to be found in President Wilson's Fourteen Points.

But our concern here is not to discuss these two modes of relationship. It is rather to note a change and an opportunity. Our new technologies of communications and transport, the multiplying channels of international contact and their new reach and speed, have made communications to peoples at large a fact of international life, bypassing the channels of statist diplomacy. Relations with our allies, with our adversaries, with those peoples who wish to be neither, all are inevitably shaped in new, intimate, and immediate ways by the ready—almost automatic—flow of ideas and information worldwide. At least in this area we can now escape the conflict between the two forms of diplomacy, and synthesize them as harmonious rather than as divergent approaches to our international relationships.

This is precisely why the Soviet Union seeks to limit such open communications, since this atmosphere directly challenges its monopoly of truth. Yet the Soviet leadership, governing a powerful industrialized nation, must pay some attention to world public opinion; it cannot totally ignore accepted modes of international contact and conduct. The tension produced by this dilemma is aggravated because the Soviet government itself has encouraged and succeeded in producing a more highly educated population with extensive mass media. Increasingly, to try to control all communications in such a population is to make nonsense of Soviet pretensions in the international sphere.

The technical apparatus and the general norm of open communications among the peoples of the world create an opportunity to complement statist diplomacy. International broadcasting can play an essential role in this process, for it alone permits something resembling the free intercourse of peoples, as distinct from governments. It therefore acquires a special opportunity and special responsibility to concern itself with humane and democratic values and help to meet the need affirmed in the Helsinki Declaration for "an ever wider knowledge and understanding of the various aspects of life in other participating states." There is thus available a two-track or dual diplomacy, with two mutually reenforcing aspects.

Deterrents and Disparities

Our East-West diplomacy aims at negotiations, understandings, agreements with the Soviet Union—fostering the evolution of a safer relationship with a superpower possessing the military capacity to destroy much of mankind. To the extent that we can reduce areas of friction and limit the central armaments of the superpowers, we strengthen and stabilize the deterrent strategy seeking to prevent nuclear or conventional war. The Limited Test Ban Treaty, the Berlin accords, the Anti-Ballistic Missile Treaty, the Interim Agreement on offensive missiles—these have been painstakingly negotiated and solemnly formalized. And we have mobilized our most sophisticated technology in elaborate systems monitoring compliance with these agreements.

But on the communist side, the ultimate sanctity of these agreements depends on the attitudes and perceptions of the few men who rule the Soviet Union, and the true terrain of deterrence is their state of mind. No small part of the $93 billion of our 1976 defense budget, and the $51 billion expended on defense by our NATO allies, is devoted to affecting states of mind.

Although power is concentrated in a relatively few people in the Soviet Union, their views and attitudes—the perceptions that deter dangerous decisions and establish the goals and responses of their diplomacy—cannot be totally isolated from the public state of mind. Even in the most rigidly totalitarian society, what the people know, think, and feel is of some importance. There could be no more eloquent testimony to the latent power of public opinion than the immense apparatus for indoctrination and control of access to information that operates in the Soviet Union and the Warsaw Pact countries. What the public knows and understands inevitably helps to shape the choices of their leaders, to impose rationality and restraint, or to give rein to reck- lessness and aggression. W. Averell Harriman, our wartime ambas- sador to the Soviet Union, testifying on behalf of funds for RL and RFE, emphasized that he viewed the flow of information to the peoples of the communist countries as

> one of the most important protections to our national security. An ignorant people will be easily led by the Kremlin and other Politburos. An informed opinion is far more difficult to dominate [Report, 1973].

The immense effort of communist governments to mold the minds of their citizens, and the vast military expenditures of the West to affect the perceptions and decisions of Soviet leaders, underscore a striking asymmetry in our concept of deterrence. It is astonishing that so little of our effort and resources is devoted to developing a communications diplomacy to reach the peoples of Eastern Europe and the Soviet Union.

We have not only expended huge sums of money to create our vast deterrent weapons systems and our NATO alliance strategy, but to develop and perfect them we have mobilized great intellec- tual resources, in and out of government. Experts from the uni- versities, industry, and the scientific community have joined in formulating and analyzing the concept of deterrence, and in pro- ducing armaments. Its stability, its effectiveness, its dangers— even its morality—are constantly assessed and debated.

But what comparable effort is there which, coherently and thoughtfully tackles the problem of creating a body of informed public opinion, without which deterrence must depend solely on the unfettered decisions of a few leaders?

The people of Eastern Europe and the Soviet Union have as great a stake as we in ensuring effective deterrence. Their role in the political process can tip the scale between its success—or its hideous failure. And, although armaments may deter, over the long run a peaceful and constructive relationship must have its basis in the concerns, views, preoccupations, and understanding of the public. To go beyond the negative goal of averting war to the positive goal of a changed and hopeful international political climate requires a new concept of communications diplomacy with the Soviet Union and Eastern Europe. Such a concept can serve the goals both of our formal policy and diplomacy, concerned for security, and of the larger diplomacy of democratic peoples, concerned for human rights and freedoms.

A Policy for the Future

In forging a new broadcast diplomacy we must first consider the prospective audiences. Beyond the dissidents and the producers of *samizdat* there are other important groups who, not only from ethnic or national traditions, see Soviet authority as alien or hostile.

(1) There are those who have been called "the politically curious." They may not have any defined dissenting views nor be in any way activists, nor may they be intellectuals. But they are intelligently aware that officially-controlled sources provide a very incomplete and self-serving version of the world. They look for other interpretations—though they may not accept them. Their curiosity may not be instrumental in seeking change; it nevertheless makes them couriers and catalysts of a wider world.

(2) There are the managerial and professional classes: the executives and planners in scientific, industrial, agricultural, engineering—and military—enterprises, who do not dissent from the system, but seek its amendment. Some are idealists, some careerists, some concerned primarily with efficiency. They may have misgivings about a stagnant economic system, or see excessive resource allocation to the military field as frustrating the promises of Marxism, or look for a more flexibly educated and trained manpower base. Increasingly, the power and performance of any modern industrialized state depends on the efficiency of this group, and they cannot be efficient without an increasingly so-

phisticated and extensive supply of undistorted information, often from abroad. However carefully rationed and controlled, such information serves—at the very least—to nourish pragmatic doubts tending to curb dogmatic decisions. Though the pressure of broad public opinion may operate only slowly and diffusely, the views of this group, vital to the machinery of state and society, inevitably exert a more direct influence on the ruling elite.

(3) Finally, there is the younger generation—a constantly replenished reservoir of challenge to any status quo—who will determine the nature of the Soviet Union in the future.

To reach these audiences more effectively, the United States needs a far clearer concept, and must develop a better organizational structure to implement it, and must devote a far greater share of resources to the task. The new concept of communications diplomacy must entail recognition of the need to allocate much greater resources to a policy to build a broadcast bridge to the minds of Soviet and Eastern European citizens. The new approach must recognize that so long as Soviet leaders insist that détente and coexistence mean intensifying the ideological struggle and tightening censorship, the United States must openly and purposefully aim at bringing ideas and information to these closed societies. If leaders who control vast armies and nuclear weapons can manipulate public opinion and make completely unfettered decisions, there must be increased danger of war through miscalculation or irrationality.

Such a new concept for broadcast diplomacy would mean re-educating some of our old-fashioned diplomats who do not appreciate the need for a dynamic effort on a people-to-people level. It also entails reassessing the present organizational structure and responsibilities of VOA. As part of USIA, VOA is generally perceived abroad as an integral and official part of the United States government—far more so than the BBC or DW. And indeed, by virtue of its present placement, VOA must necessarily often directly serve immediate foreign policy interests. But in today's world the United States has to speak freely and forthrightly on human rights and values.

In fact, with this end in view, a United States House of Representatives International Relations Subcommittee report has suggested that RL and RFE, under the Board for International Broadcasting, should extend their broadcasts to other authoritarian areas of the world. This author, however, believes such a task would be better performed, not by an expanded worldwide role for the RL and RFE corporations under the Board for International Broadcasting, but rather by a reorganization of VOA under its own board, like the BBC, where it would be buffered from the oversensitized restraints and requirements of official foreign policy. VOA, not RL and RFE, has available the worldwide transmission network and the language skills. These, of course, should be augmented to carry out an enlarged program to other totalitarian and authoritarian areas.

As for Eastern Europe and the Soviet Union, receptive audiences offer expanded programming opportunities for both a revitalized VOA and RL and RFE. As in the past, VOA should deal primarily with worldwide news and American foreign policy and culture, with RL and RFE striving to take the place of a home service. It has often been suggested, with considerable merit, that Radio Liberty and Radio Free Europe be renamed to express their present functions more exactly. (One suggestion has been "Radio Human Rights.")

Full achievement of a revitalized effort depends upon an extensive modernization of all these American radios. Since this would be a multimillion dollar effort, all facilities whether modernized, built, or leased overseas from foreign governments should be the subject of a joint RL and RFE plan with VOA. This plan might best be devised and executed by an executive committee of the Board for the VOA and the Board for International Broadcasting, with ownership of all future facilities for RL and RFE vested in BIB. The latter move would:

(1) enable government-wide contingency plans to be made in case of loss of any facility;

(2) it would often save foreign import taxes formerly levied on RL or RFE transmitters;

(3) it would help foreign governments to ward off communist pressure; and

(4) it would permit trade-off rentals among other Western governmental broadcasters, like the current arrangement where VOA uses the BBC Wolferton facilities.

New modes of cooperation could also serve the United States well in times of crisis, or even in extraordinary situations where public information is essential to reinforce diplomatic action. Twice VOA has massed transmitters in saturation broadcasts to surmount jamming. In November 1961, Edward R. Murrow, then director of USIA, ordered a massive effort to tell the Soviet bloc of world reaction to the secret and extensive high-megaton nuclear testing of the Soviet Union. A year later in the Cuban missile crisis, 52 transmitters on 80 different frequencies broadcast news of the American actions and position for over eight hours (Whitton and Larson, 1964: 50-51; Sorenson, 1968: 205-206). The coordinated use of all United States' financed transmitters—VOA, RL, and RFE—is an important consideration for the future strategy of international broadcasting; such a step forward by the United States could also lead to concerted efforts at times of crisis by other broadcasters of Western Europe.

The Importance of Western Europe

To strengthen Western European international broadcasting to the East, the countries involved must be given a better appreciation of the need for such broadcasts and the complementary roles played by the various Western broadcasters. Atlantic community groups, such as the North Atlantic Assembly of Parliamentarians, the North Atlantic Treaty Association, the Atlantic Council, and the Atlantic Institute, can explain the new role of international broadcasting in East-West relations. This would encourage the British Parliament to give stronger financial support to the BBC's External Service. It can also help to sustain the new support for Deutsche Welle which, unlike other Western broadcasters, has already moved into a $70 million modernization program with the installation of nine new 500-kilowatt transmitters at Wertachtal, West Germany. These efforts can also show the importance of Sweden's and Canada's broadcasts to communist countries.

It was, of course, highly regrettable that France's Foreign Broadcasting Service dropped all Eastern broadcasts at the end of 1974, especially in view of the richness of French culture and its international ties. The recent announcement that France plans to resume broadcasting to Eastern Europe is welcome. A rejuvenated and coherent, cooperative international broadcasting concept would benefit greatly from French participation. The cultural talent available in Paris, both from Russia and France, makes possible richly diversified programming in expanded French international broadcasts.

All these complementary endeavors have a role to play in bringing information about Western nations to Eastern nations to help counteract the suspicions and isolation that lead to mutual enmity. In addition to such national broadcasters, RFE and RL and RIAS have special significance and a unique place, since they attempt to provide a surrogate home service for Eastern Europe and the Soviet Union. They alone command the depth of linguistic and research capabilities that enable them to devote over half their programming to internal developments in the audience countries. The presidential commission chaired by Dr. Milton Eisenhower recommended that the RL and RFE broadcasts should be funded only by the United States; any kind of "multilateral broadcasting force" would be far too cumbersome to work, and the resultant bureaucracy might stifle the radios with censorship. But Western European countries can and do play other important and often essential roles in making these efforts possible.

The most obvious is providing the sites for transmitters, which at present must be about 2,000 miles away from the audience areas to be effective. The present state of technology does not allow direct effective transmission to Eastern Europe from the United States; it would take the power of a nuclear plant to produce the necessary three signal hops to the ionosphere and back. Satellites are good for relays, provided the satellite is within reach, or can "see" both earth stations at the same time. But they cannot successfully transmit to domestic radio sets. The United States educational satellite used to aid India transmits only to expensive receivers for instruction of large groups.

Thus the mainland of Western Europe—the transmission points in Spain, Portugal, West Germany, and any new sites needed as a result of modernization—remains essential to these U.S. funded broadcasts. It is important for these host countries to understand that these efforts are a joint venture in their own interest for the peace of Europe, and not simply a concession to the United

States. As already noted, this has been symbolized in Portugal by the establishment of a Portuguese corporation. As Spain seeks greater affiliation with democratic Western Europe and with the United States, her role as host to the RL transmission facilities strengthens the case for her acceptance; any Spanish decision to discontinue this role would be withdrawal from a democratic cooperative endeavor. Moreover, while RL and RFE transmitters are located only in West Germany, Portugal, and Spain, the sites of their news bureaus, audience research organizations, and program activities include also Paris, Brussels, and London, and RFE has bureaus in Rome and Bonn.

Although the Eisenhower commission wisely warned against an attempt to include European governments in financing broadcasting operations, it suggested that they might support the research. Objectives of a revitalized communications diplomacy for the West should include a joint research institute utilized by all Western broadcasters and in large part built upon present RL and RFE research. In addition to enriching the research for programming of all Western international broadcasts, such a jointly sponsored institute could help monitor for the West the outcome of Basket III.

Another area where cooperation is much needed is in the technical field, especially in joint use and swapping of transmission time, facilities, and relay stations, such as VOA's use of Wolferton in England and BBC's use of Greenville, North Carolina for its Latin American broadcasts. Modernization programs will be expensive for all governments, and cooperative endeavors will maximize capacities. There is a need for cooperation in frequency assignment; much will be determined at the 1979 International Telecommunications Union Radio Conference. A major effort must be made to expand the frequency spectrum space for high frequency broadcasters. At present, there are about 1,500 short-wave transmitters in use, so that every short-wave channel is used by more than three transmitters, leading to interference. Without an agreement to expand the frequency spectrum, the demands for more relay bases and more powerful transmitters will accelerate, and reception will deteriorate.

Broadcasting and Diplomacy

If foreign policy designs are to succeed and endure, they must be institutionalized through broad public knowledge and understanding. Lacking Soviet moves toward freer communication within its borders and with the outside world, the basis of co-existence is frail and narrow.

The greatest foreign policy designs have often failed because they have not gained, over time, public understanding based on knowledge and tested in public discussion. They remained principally the creations of diplomatists and heads of state. The concert of Europe, despite great initial and sporadic success, eventually failed because it was out of touch with the people. With Bismarck's departure, the foreign policy structure he envisaged for Germany failed because it was not understood and supported in depth in the parliament and throughout Germany. What was true in nineteenth century diplomacy applies with even greater force in the nuclear age, when some superpowers are totalitarian states. If treaties are to have sanctity, they must be known and command assent throughout the dictatorial as well as the democratic societies involved. When information and ideas cannot cross national boundaries, those frontiers remain fortified and hostile divisions. Insistence on Brezhnev's "unabating ideological war" imprisons the peoples of the Soviet Union in an implacable enmity. One noted observer of Soviet policy remarked that:

This insistence clearly has its roots in organizational politics within the Soviet system, but it presents operational problems in foreign policy, for the continued reliance of the Soviet Union upon an external ideological adversary, as a device necessary to its system of political control, sets limits in practice on the realization of its policy of "peaceful coexistence" [Shulman, 1973: 35-58].

As we have seen in this century, devil theories in politics readily become destroyers of world peace. Both the German and Japanese people were carefully indoctrinated prior to the aggressions that led to World War II. The Stalinist doctrine of capitalist encirclement and the constant stress on the menace of counterrevolutionaries are ways of maintaining those public suspicions that Soviet rulers judge so necessary for maintaining their own legitimacy. But a misinformed, suspicious public—particularly in the hands of an authoritarian regime or a dictator—is a factor in the international political process, a threat to peace and lasting accord. One need not subscribe to all of Solzhenitsyn's foreign policy positions to recognize the truth he proclaimed in his Nobel Prize lecture: "Blocking of information makes international signatures and treaties unreal; within the zone of stunned silence any treaty can easily be reinterpreted at will."

The most important question here is whether coexistence can be fruitful and durable if one of the systems permits only its own version of the past, the present, and the future to circulate freely. In a world growing steadily richer in its capacity for violence and steadily poorer in the resources for life, both sides must have common access to the stark facts. The two great nuclear powers share common dangers and a common responsibility to help meet the issues of an increasingly interdependent world. Détente may not grow into *entente,* but there needs to be some common understanding of the doubts, fears, and aspirations that all people must confront and share. If ideas and facts in one country must conform to doctrinally fixed criteria of acceptability, if intellectual life is restricted to an authorized set of premises, subjects, and conclusions, détente is limited and the dangers of a divided world are perpetuated.

In a dangerously divided, nuclear-armed world, people must share—with each other and among themselves—some basic measure of common facts and ideas if they are to make common cause. International radio today can serve to undermine an adversary relationship set up by one side with a closed system. It must show the world in its complexity and reveal people to each other and peoples to themselves.

REFERENCES

BARGHOORN, F. C. (1964) Soviet Foreign Propaganda. Princeton: Princeton Univ. Press.

BARRETT, E. W. (1953) Truth is Our Weapon. New York: Funk & Wagnalls.

BBC Handbook (1974, 1975, 1976) London: British Broadcasting Corp.

Board for International Broadcasting [BIB] (1975) 2nd Annual Report (October) Washington, D.C.: U.S. Gov. Print. Office.

BROWN, A. (1976) Director of Programming, RCI, Montreal. Telephone interview (April 7).

BROWN, J.A.C. (1965) Techniques of Persuasion. Baltimore: Penguin.

CODDING, G. A. (1959) Broadcasting Without Barriers. New York: UNESCO.

Commission on the Organization of the Government for the Conduct of Foreign Policy (1975) Report (June) Washington, D.C.: U.S. Gov. Print. Office.

Comptroller General of the U.S. (1972) "U.S. government monies provided to Radio Free Europe and Radio Liberty." Report to the Committee on Foreign Relations, U.S. Senate. Washington, D.C.: U.S. Gov. Print. Office.

de SOLA POOL, Ithiel (1965) Address in Workshop on Communications with the Peoples of the U.S.S.R. (November 19) N.Y.U. School of Education, Dept. of Communications in Education, and Radio Liberty Committee.

DEWHIRST, M. and R. FARRELL [eds.] (1973) The Soviet Censorship. Metuchen, N.J.: Scarecrow Press.

DIZARD, W. P. (1961) The Strategy of Truth. Washington, D.C.: Public Affairs Press.

ELLUL, J. (1965) Propaganda. New York: Knopf.

FRASER, L. (1957) Propaganda. London: Oxford Univ. Press.

GREENE, Sir H. (1969) The Third Floor Front. London: Bodley Head.

HALE, J. (1975) Radio Power. London: Paul Elek.

HANNAH (formerly Hollander), G. D. (1972) Soviet Political Indoctrination. New York: Praeger.

KATZ, Z. (1974) "Some basic processes in Soviet society." Center for International Studies, M.I.T.

LARRABEE, F. S. (1976) "Soviet attitudes and policy towards 'Basket Three' since Helsinki." Radio Liberty Research Report, Munich (March 15).

LEONHARD, W. (1973) "The domestic politics of the new Soviet foreign policy." Foreign Affairs (October): 59-74.

Library of Congress, Congressional Research Services (1972a) Radio Liberty—A Study of its Origins, Structure, Policy, Programming and Effectiveness (February).

——— (1972b) Radio Free Europe—A Survey and Analysis (February).

LISANN, M. (1975) Broadcasting to the Soviet Union: International Politics and Radio. New York: Praeger.

MEDVEDEV, R. (1971) Let History Judge: The Origins and Consequences of Stalinism. New York: Random House.

MEDVEDEV, Z. A. (1973) Ten Years after Ivan Denisovich. New York: Random House.

MICKIEWICZ. [ed.] (1973) Handbook of Soviet Social Science Data. New York: Free Press.

N.Y. Times (1975) 8 October.

OSNOS, P. (1976) "East Germans spur efforts to counter TV image of west." Washington Post (May 28).

PANFILOV, A. F. (1967) U.S. Radio in Psychological Warfare. Moscow: International Relations Pub.

PAULU, B. (1974) Radio and Television Broadcasting in Eastern Europe. Minneapolis: Univ. of Minnesota Press.

POND, E. (1976) "Soviets blast U.S. radio 'intrusion.'" Christian Science Monitor (January 14).

Pravda (1976) 13 January. Cited in E. Pond "Soviets blast U.S. radio 'intrusion.'" Christian Science Monitor (January 14, 1976).

Pravda (1976) February 26.

Radio Canada International (1976) Program Schedule No. 95 (May 2) Montreal.

––– (1973) General Information (December) Montreal.

Report of the Presidential Study Commission on International Radio Broadcasting (1973) "The right to know." Washington, D.C.: U.S. Gov. Print. Office.

ROBERTS, W. R. (1973) Tito, Mikailovic and the Allies, 1941-1945. New Brunswick, N.J.: Rutgers Univ. Press.

SALISBURY, H. E. [ed.] (1974) Sakharov Speaks. New York: Random House.

SHULMAN, M. D. (1973) "Toward a Western philosphy of coexistence." Foreign Affairs (October): 35-58.

SORENSEN, T. C. (1968) The Word War: The Story of American Propaganda. New York: Harper & Row.

STANTON, F. and panel (1975) International Information, Education, and Cultural Relations: Recommendations for the Future. Washington, D.C.: Center for Strategic and International Studies.

SZPORLUK, R. (1976) The Influence of East Europe and the Soviet West on the USSR. New York: Praeger.

THOMPSON, C.A.H. (1948) Overseas Information Services of the United States Government. Washington, D.C.: Brookings Institution.

TOKES, R. L. [ed.] (1975) Dissent in the USSR. Baltimore and London: Johns Hopkins Univ. Press.

ULAM, A. (1973) Stalin: The Man and His Era. New York: Viking.

U.S. Congress, Public Law 93-129 (1973) "Board for International Broadcasting Act of 1973," 93rd Congress, S. 1914 (October 19).

90

U.S. Department of State, Bureau of Public Affairs (1975) Conference on Security and Co-operation in Europe: Final Act, Helsinki (August).

U.S. Information Agency, Office of Research and Assessment (1973) External Information and Cultural Relations Programs, Federal Republic of Germany.

WHITTON, J. B. and A. Larson (1964) Propaganda. New York: Oceana.

World Radio TV Handbook (1976) J. M. Frost [ed.] New York: Billboard.

Take advantage of the special 20 percent discount on CSIS publications available only to *Washington Paper* subscribers!

If you subscribe to the *Washington Papers*, you are entitled to a special 20 percent discount on CSIS books, research monographs, and conference reports. In the quality tradition of the *Washington Papers*, these publications are available by mail from Georgetown University Center for Strategic and International Studies, 1800 K St., N.W., Washington, D.C. 20006. To qualify for the discount, you must include payment with your order and indicate that you are a *Washington Paper* subscriber. Please add 30c per publication for book-rate postage and handling.

	List Price	Discount Price
1. **World Power Assessment**, Ray S. Cline (1975); maps charts, tables, 173 pp.	$6.95	$5.55
2. **U.S./European Economic Cooperation in Military and Civil Technology**, Thomas A. Callaghan Jr. (1975); 126 pp.	$3.95	$3.15
3. **Foreign Policy Contingencies: The Next Five Years**, Edward Luttwak, Rapporteur (1975); 21 pp.	$1.95	$1.75
4. **Soviet Arms Aid in the Middle East**, Roger F. Pajak (1976); 45 pp.	$3.00	$2.60
5. **Communist Indochina: Problems, Policies and Superpower Involvement**, Joseph C. Kun (1976); 38 pp.	$3.00	$2.60
6. **Indonesia's Oil**, Sevinc Carlson (1976); maps, charts, tables, bibliography, 89 pp.	$3.95	$3.15
7. **A U.S. Guarantee for Israel?**, Mark A. Bruzonsky (1976); 62 pp.	$3.00	$2.60
8. **Armed Forces in the NATO Alliance**, Ulrich de Maiziere (1976); 48 pp.	$3.00	$2.60
9. **Europe, Japan, Canada, and the U.S.: The Interaction of Economic, Political, and Security Issues** (Third Quadrangular Conference), Edward Luttwak, Rapporteur (1976); 37 pp.	$3.95	$3.15
10. **The Political Stability of Italy**, Endre Marton, Rapporteur (1976); 67 pp.	$4.00	$3.20
11. **China Diary**, Harlan Cleveland (1976); 50 pp.	$3.00	$2.60
12. **The Soviet Union: Society and Policy** (Williamsburg Conference III), Endre Marton, Rapporteur (1976); 48 pp.	$3.00	$2.60

CSIS National Energy Seminar Reports, Francis C. Murray, Ed.

	List Price	Discount Price
13. **Deregulation of Natural Gas** (1976); 57 pp.	$3.00	$2.60
14. **The Energy Independence Authority** (1976); 56 pp.	$3.00	$2.60
15. **Divestiture: The Pros and Cons** (1976); 68 pp.	$3.00	$2.60